Transistor Basics

A Short Course

REVISED SECOND EDITION

Transistor Basics

A Short Course

REVISED SECOND EDITION

GEORGE C. STANLEY, JR.
Hewlett-Packard Company

HAYDEN BOOK COMPANY, INC.
Rochelle Park, New Jersey

Library of Congress Cataloging in Publication Data

Stanley, George C
Transistor basics, a short course.

Includes index.
1. Transistors. I. Title.
TK7871.9.S7 1975 621.3815'28 74-34241
ISBN 0-8104-5866-7

Printed in the United States of America

6 7 8 9 PRINTING

81 82 83 84 YEAR

Preface to Second Edition

The wide acceptance of the first edition by both service and engineering personnel made me realize there is a great need for concise practical information on transistors, transistorized circuits, and troubleshooting techniques. As a result, this second edition is expanded in these directions and includes considerably more material on troubleshooting PNP and NPN transistors as well as FET's. Both of these new sections include useful testers for checking semiconductor devices. Also included is new material and a simplified analysis relating to the emitter follower. The result of these additions should make this second edition more useful to those of us dealing with electronics on a regular basis.

GEORGE C. STANLEY, JR.

Preface to First Edition

This book should prove useful to persons with many different backgrounds. Anyone having a superficial knowledge of transistors, for example, will find that the book transforms this knowledge into a working capability. The technician or engineer engaged in the maintenance of transistorized equipment will find the book to be an excellent source of practical information. Finally, the technical institute or college student taking a first course in transistors will find that this material forms a truly worthwhile supplement to any textbook he might be using.

Perhaps the most useful feature of this book is the development of a simplified approach to transistor circuit analysis. By using this approach, the reader will be able to determine by inspection the values of such quantities as voltage gain and current gain with an accuracy of ± 10 percent. Accuracies of this order are adequate for almost any practical situation.

Because the several approximations used in developing the simplified analysis technique are, of necessity, based on the operating characteristics of present-day transistors, slight changes are bound to occur in coming years. To keep the book from becoming obsolete at that time, a table of exact expressions plus a set of rule-of-thumb formulas are presented in appendices.

In the preparation of this material many helpful suggestions were offered by Ben O. Lange. These suggestions were the result of a course in transistors he taught at Foothill College, Los Altos, California. Special thanks also go to Pat Lynch for her editorial help and to Ernie Bennett and Vera Vacek for their artwork contributions.

GEORGE C. STANLEY, JR.

Contents

Symbols

A_e	A-c voltage gain of transistor circuit
A_i	A-c current gain of transistor circuit
A_p	Power gain for transistorized circuit
β	For practical purposes, the same as h_{fe}
β_{DC}	For practical purposes, the same as h_{FE}
g_m	Transconductance of field-effect transistor
h_{FE}	Transistor d-c forward current gain for common-emitter configuration
h_{fe}	Transistor a-c short-circuit forward current gain for common-emitter configuration
h_{ib}	Transistor a-c short-circuit input impedance for common-base configuration
h_{ie}	Transistor a-c short-circuit input impedance for common-emitter configuration
h_{oe}	Transistor a-c short-circuit output admittance for common-emitter configuration
h_{re}	Transistor a-c short-circuit reverse voltage gain for common-emitter configuration
i_b	Instantaneous a-c value of base current
i_c	Instantaneous a-c value of collector current
I_{CBO}	Collector-base leakage current (with emitter open)
I_{CO}	Same as I_{CBO}
I_{DSS}	Drain-to-source current with zero volts between gate and source

$\mathbf{I}_{GSS}$	Leakage current in an FET between gate and source with drain and source shorted together
$\mathbf{r}_e$	A-c emitter resistance in *"r"* parameter notation
$\mathbf{R}_{DS}$	Drain source (channel) resistance in an FET
$\mathbf{R}_i$	A-c input impedance of transistor circuit
$\mathbf{R}_o$	A-c output impedance of transistor circuit
S	Current stability factor
T.C.	Temperature coefficient
$\mathbf{V}_{CC}$	D-c value of supply voltage

CHAPTER 1

Introduction

Transistors started to appear on the market about 1951, but the performance capabilities of the early devices, in comparison to vacuum tubes, left much to be desired. The technology has improved to such an extent, however, that the transistor has now replaced the vacuum tube in most applications and this trend continues. Thus, an understanding of transistors is essential to the modern electronics technician and engineer.

In common with other technologies, certain terminology is associated with transistors and an understanding of this terminology is necessary before we can proceed further.

1.1 Transistor Terminology

Electron: This small negatively charged particle is most important in semiconductor operation. We are interested in the electron in one of two states: (1) *bound* in the lattice framework of a material, or (2) *free* of the lattice and able to flow through the material as current.

Hole: In a semiconductor material at room temperature, a crystal lattice structure always has imperfections where electrons should be. If this were not so, and all electrons were in place, the lattice would be perfect and would be electrically neutral. However, when an electron is missing there is a localized positive charge at the point of imperfection. It is this *imperfection in the lattice, having a positive charge just equal to that of the electron, that is called a hole.* These imperfections (holes) can be formed by heat energy, light energy, electric fields, random behavior of the lattice, or, as we will see, by doping.

Doping: The fundamental process of doping introduces impurities into the lattice structure that are compatible with the existing lattice. However, the impurity atoms differ slightly from the lattice atoms in that they either have one extra or one less electron in their valence band (i.e., the valence electrons that form the lattice). Thus, through doping, we can create a lattice that has a few holes or a lattice that is completely full with free electrons left over. If the material has been doped to have extra electrons it is called N-doped. If it is doped to have holes, it is called P-doped.

In the above discussion, one item has not been mentioned: even after doping, the material is *still* electrically neutral. This may not be immediately apparent when we are doping with N or P materials. The answer is that the doping materials are themselves electrically neutral. When arsenic is used to N dope a semiconductor, for example, the negative charge of the extra electrons is just counterbalanced by the positive charge existing within the nucleus of the arsenic atoms.

Majority Carrier: This is most easily explained with an example. A free electron in an N-doped semiconductor is a majority carrier.

Minority Carrier: Using the same example as above, an N-doped semiconductor, a hole in the lattice structure is a minority carrier. This hole most likely would be created by the thermal energy at room temperature breaking a bound electron loose from the valence band leaving a hole. In summary, an electron in N material is a *majority* carrier, an electron in P material is a *minority* carrier, a hole in P material is a *majority* carrier, and a hole in N material is a *minority* carrier.

Stored Minority Carriers: All previous terms are necessary to adequately explain this term. It is either stored minority carriers or the lack of them that dominates the behavior of most diodes including step recovery diodes, PIN (positive-intrinsic-negative) diodes, hot carrier devices, high conductance diodes, and just plain PN diodes.

An example of a stored minority carrier is an electron finding itself in P material when a diode is reverse biased. Simply put, it is the wrong particle in the wrong place at the wrong time.

One way in which stored minority carriers can become troublesome is shown in the simple transistor circuit of Fig. 1-1. Here the positive-going waveform applied to the

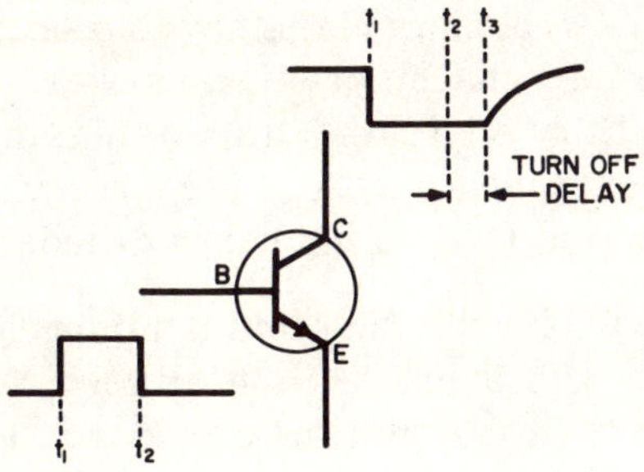

Fig. 1-1. The effects of minority-carrier storage.

base of the transistor at time t_1 turns on the emitter-base diode and pulls many electrons into the base region. Actually, more electrons are drawn into the base than can continue into the collector.

This condition is about the same as that which occurs when there is heavy air traffic into Los Angeles, Chicago, or New York City on a late Sunday afternoon. More aircraft are entering these areas than can land, and the airplanes stack up over the cities. This is just what happens in the base of our NPN transistor. Because of the high forward bias, many electrons move from the emitter into the base; so many in fact that they stack up in the base waiting to get into the collector. These electrons are majority carriers in the emitter, but when they move into the P-type material base they become minority carriers. If these carriers have to store in the base waiting to get into the collector, they are *stored minority carriers.* When the base bias is finally turned off at time t_2, these stored minority carriers (just like the stacked airplanes) move into the collector. After enough of them have moved into the collector, the transistor begins to move out of saturation (time t_3), and starts to turn off. Obviously, the output pulse is not a true representation of the input pulse and the difference (turn-off delay) is due to the stored minority carriers (electrons) in the base (P) material. The behavior of stored minority carriers is the key to diode performance. To produce certain characteristics, carrier storage can be minimized (with a switching diode), maximized (with a PIN diode), controlled (with a step recovery diode), or virtually eliminated (with a hot carrier diode).

Lifetime: (τ) This is the time a minority carrier survives; i.e., it is the average time an electron can last in P material and the average time a hole can last in N material. Being more specific, it is the period of time it takes for 63 percent of the minority carriers injected across a junction to recombine. Actual values are highly dependent upon the amount of doping and can easily vary by a factor of 1000. Lifetime is important as it determines how fast a diode can turn off.

PN Junction: A PN junction is created when N-doped and P-doped materials are tightly bonded together into a single crystalline structure. Immediately after contact is made and right along the junction, electrons from the N material start falling into all the nearby holes in the P material. This action depletes the region of free carriers right at the junction and forms a *depletion region* as shown in Fig. 1-2. Note that in the region of the junction the N material acquires a slight positive charge and the P material a slight negative charge. This junction-charge difference creates a *barrier potential* and comes about because both P and N sections were neutral before joining.

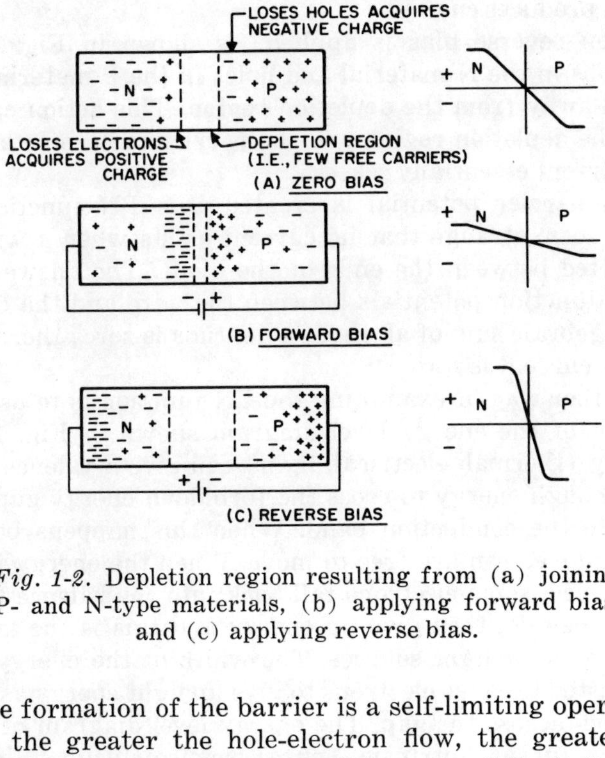

Fig. 1-2. Depletion region resulting from (a) joining P- and N-type materials, (b) applying forward bias, and (c) applying reverse bias.

The formation of the barrier is a self-limiting operation, since the greater the hole-electron flow, the greater the junction potential difference and the more difficult it is for an electron or hole to cross the barrier.

In forming the barrier, the electrons can be thought of

as rolling down the potential hill, and the holes as floating up the hill. This can be visualized if you think of a water filled tube with marbles at one end and bubbles at the other. Start tipping the tube (formation of barrier) and the marbles (electrons) roll down the hill while the air bubbles (holes) float up the hill.

When forward bias is applied to a PN diode, as shown in Fig. 1-2B, electrons in the N material and holes in the P material are pushed toward the barrier (depletion region). Two things happen: the depletion region is made smaller and the barrier height is reduced. This combination of events produces current.

When reverse bias is applied, as shown in Fig. 1-2C, electrons in the N material and holes in the P material are drawn away from the depletion region. This action causes both the depletion region and the barrier to become larger and current essentially ceases.

If a barrier potential is created at a PN junction, it might seem strange that no current exists when a wire is connected between the ends of the diode. The answer lies in the junction potentials between the wire and the diode. The algebraic sum of all three potentials is zero; therefore, the net current is zero.

Another way of examining the PN junction is to use the concept of the energy level diagram shown in Fig. 1-3A. Energy (thermal, electrical, light) can give a valence electron enough energy to cross the forbidden energy gap and get into the conduction band. When this happens both a hole and electron are free to move. When the energy input is removed, some electrons fall back into the valence band. When they do, they give up energy and this is the key to semiconductor light sources. The width of the energy gap is adjusted to cause electrons to give up light energy as they fall back across the gap. The energy level diagram of Fig. 1-3A is for an intrinsic (pure) semiconductor. Similar diagrams are shown for a metal, an insulator, and a doped (extrinsic) semiconductor, respectively, in Fig. 1-3B. In

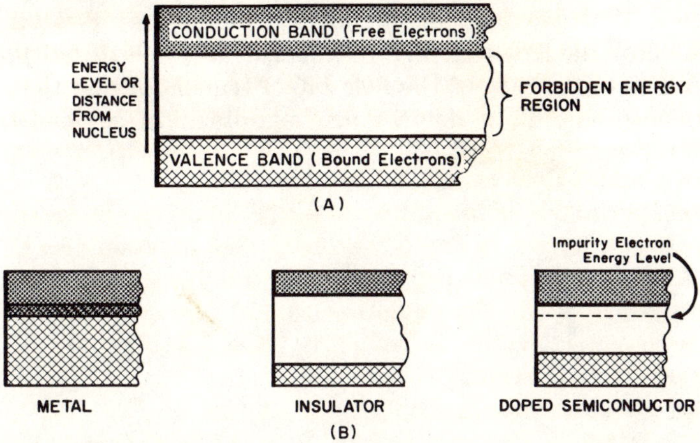

Fig. 1-3. Energy level diagrams for (a) an intrinsic semiconductor and (b) for a metal, an insulator, and a doped semiconductor, respectively.

the case of metal, notice that the valence and conduction bands overlap. This indicates an abundance of free charge carriers. In the insulator, very few free charges exist and the width of the forbidden region (gap) is great. In the doped semiconductor, there are fewer free charge carriers than in a metal, but more than in an insulator. Thus, the width of the forbidden energy gap is greater than that associated with a metal, but less than that of an insulator.

1.2 Summary

Special attention has been given to minority carriers and minority carrier storage. The importance of these terms cannot be overemphasized as the presence or lack of minority carriers governs the behavior of all semiconductors. Because minority-carrier storage limits frequency response and switching speed, PNP and NPN transistors may eventually decline in popularity and be replaced by other devices that are free of minority carriers. Only two semiconductor devices that are in general use operate without minority carriers; viz, the field-effect transistor and the hot carrier diode. Both devices are examined in some detail in later sections.

CHAPTER 2

Transistors and Diodes

2.1 Transistor Types

Present transistors are NPN and PNP. The schematic symbol for each type is shown in Fig. 2-1. These two types are easy to identify because the arrow always points towards the N material.

Fig. 2-1. Transistor types: (a) PNP and (b) NPN.

2.2 Conventional Current vs. Electron Flow

Before going further, it is probably best to decide if we will use conventional current, from positive to negative in the external circuit, or electron flow, from negative to positive in the external circuit. Conventional current is illustrated in Fig. 2-2A and electron flow in Fig. 2-2B. This is purely a matter of choice, but once the choice is made, we should be consistent. Conventional current has

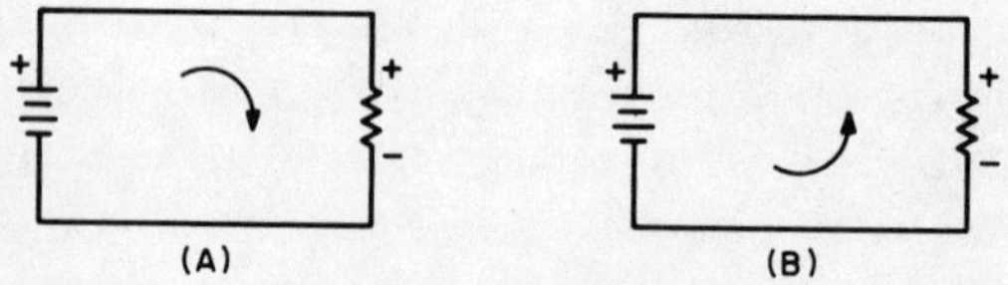

Fig. 2-2. The direction of (a) conventional current and (b) electron flow.

been the classic approach (note that it is in the direction of the arrows on the transistor symbols in Fig. 2-1). Electron flow has been adopted by the military. However, almost all transistor books have been written using conventional current and that notation is adopted here.

2.3 The PN Diode Junction

The heart of a transistor is the PN diode junction between the transistor emitter and base. To better understand transistor operation, it may be helpful to examine silicon and germanium diodes. The V-I characteristic of each type is shown in Fig. 2-3.

Notice that neither germanium nor silicon diodes conduct appreciable current until a certain minimum forward bias is reached; approximately 0.15 to 0.2 volt for germanium and 0.6 to 0.7 volt for silicon. Using this information and measuring a transistor's emitter-base potential can tell a great deal about transistor behavior. If, for example, a

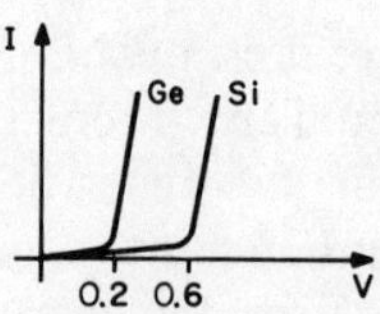

Fig. 2-3. V-I characteristics for germanium (Ge) and silicon (Si) diodes.

germanium transistor had 0.15-volt forward bias on its emitter-base junction, it would be lightly biased in the forward direction and probably draw 1 to 2 mA of collector current. In a similar manner, 0.6-volt forward bias across a silicon transistor emitter-base junction would turn the transistor on lightly.

2.4 Temperature Effects

A germanium diode, with a forward-biasing voltage applied, is shown in Fig. 2-4A. Under normal conditions, the voltage drop appearing across the diode is approximately 0.15 volt.

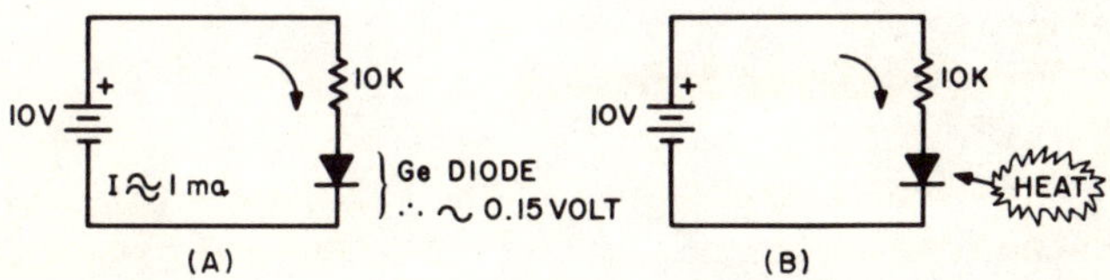

Fig. 2-4. Diode behavior with temperature variations.

When heat is applied to the diode, as shown in Fig. 2-4B, the diode tries to conduct more but the circuit current, for all practical purposes, is determined by the 10K resistor. Since the resistance of the diode decreases, however, the voltage appearing across its terminals also decreases.

The application of heat to any forward-biased diode or transistor produces the same effect. The resistance of the diode or transistor decreases as it tries to conduct more. When voltage decreases with increasing temperature, the circuit is said to exhibit a *negative temperature coefficient of resistance.*

In the case of a reverse-biased diode, the application of heat again makes the diode try to turn on harder. Since the diode is reverse-biased, however, its impedance is very high and this impedance acts to limit circuit current. As the temperature increases, current through the diode increases, but the diode impedance is only slightly reduced. As a result, the voltage across the diode goes up, and the circuit is said to exhibit a *positive* temperature coefficient of resistance.

CHAPTER 3

Leakage Current, Stabilization, and Biasing

We have seen how heat affects diode performance. Heat has a similar effect on transistors. In both cases, the heat tries to make the device conduct more (turn on harder). Since heat plays such a large role in transistor performance, it is worthwhile to examine this effect and what can be done to minimize it.

3.1 I_{CO} (Leakage Current)

A common-emitter amplifier, using a PNP germanium transistor and conventional biasing, is shown in Fig. 3-1. The emitter-base junction is forward biased and a reverse bias is applied to the collector-base junction. The biasing conditions between the collector and base are further illustrated in Fig. 3-2. Under these conditions, a leakage (reverse) current, termed I_{CBO} or, simply, I_{CO}, exists from collector to base. This current is indicated in the diode characteristics of Fig. 3-3. A second leakage current, termed I_{CEO}, exists between the collector and emitter. Current I_{CEO} is larger than I_{CBO} by a factor about equal to the current gain of the transistor.

The simplest form of biasing is shown in Fig. 3-4. Since

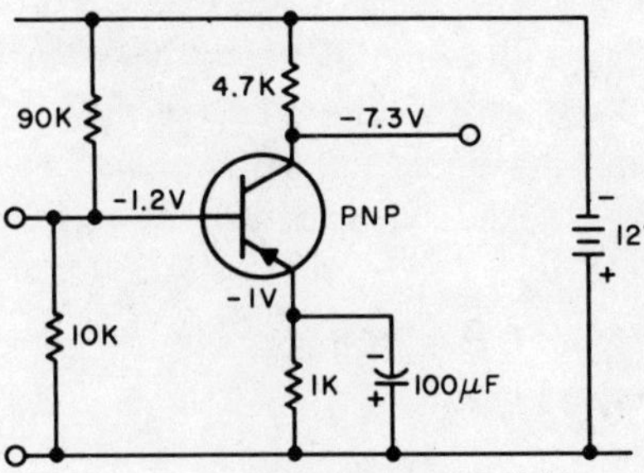

Fig. 3-1. Common-emitter amplifier.

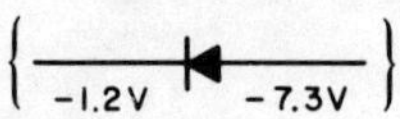

Fig. 3-2. Bias conditions.

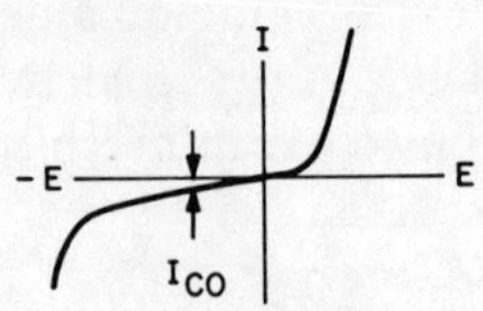

Fig. 3-3. PN diode characteristics.

the emitter-to-base junction is forward biased, the transistor is conducting.

Remember the base-collector is a reverse-biased PN junction and I_{CO} is the reverse-biased leakage current, which increases as temperature increases (a law of nature). Where does this current go? As shown in Fig. 3-5 current I_{CO} must pass through the forward-biased emitter-base junction and then across the reverse-biased base-collector junction. This extra current through the emitter-base junction acts just like a signal current. The transistor turns on harder and this increases the temperature of the transistor due to power dissipation. This extra heat causes

I_{CO} to increase still more and the process noted above is repeated. There is now *positive thermal feedback,* which often continues until the transistor is destroyed. This destructive condition is called *thermal runaway.* The solution to this problem is to prevent the emitter-base bias from

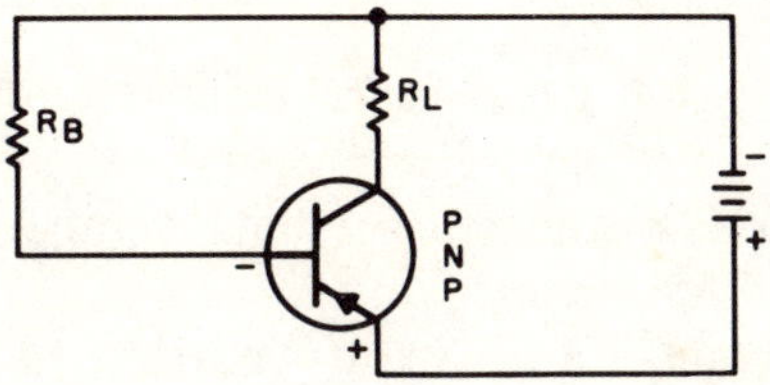

Fig. 3-4. Simple bias circuit.

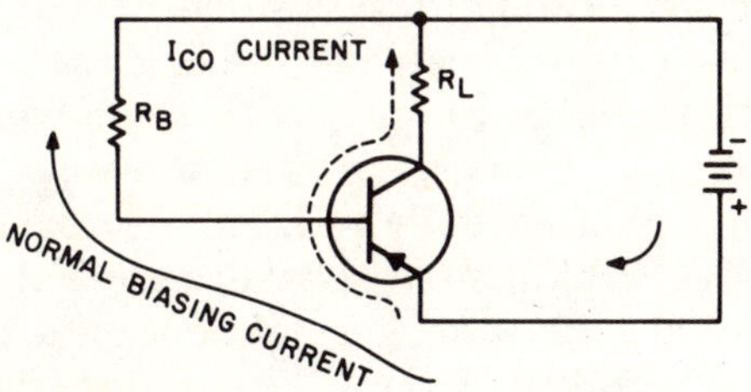

Fig. 3-5. Leakage current path.

changing because of changes in I_{CO}. This can be accomplished by having base bias independent of collector bias; i.e., by using a second battery.

Alternative circuit arrangements are shown in Fig. 3-6. Unfortunately, both arrangements are wasteful in terms of power consumption and some compromise must be reached.

3.2 Stabilization

Two important questions in transistor operation are: (1) Is the circuit temperature stable? (2) How can the proper

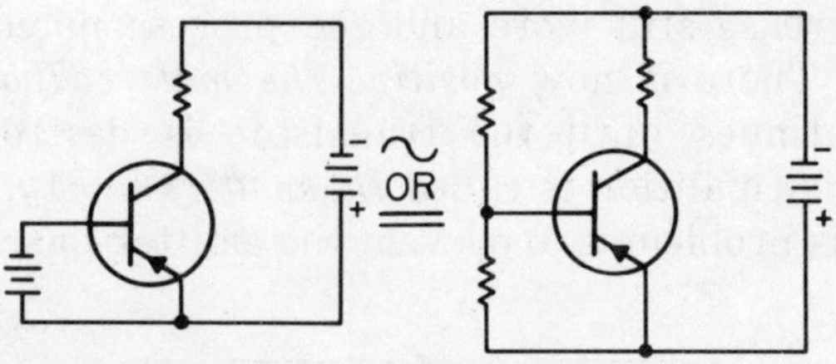

Fig. 3-6. Alternative biasing circuits.

biasing network be selected for both temperature stability and the desired operating point?

Obtaining exact stability information has always been rather difficult. To determine if a device is thermally stable is best handled by graphical analysis but this is a rather time-consuming process. Fortunately, two factors brighten the picture considerably. First, a great deal of information about stability can be obtained by using the so-called *current stability factor* S. Second, with the advent of silicon surface-passivated transistors, the problems of I_{CBO} and thermal runaway are greatly reduced. For example, silicon surface-passivated transistors have a typical I_{CBO} that is 1/100 or 1/1000 of I_{CBO} for germanium units.

The current stability factor:

$$S \equiv \frac{\Delta I_C}{\Delta I_{CBO}}$$

is a measure of how much I_C varies when I_{CBO} varies. Since we do not want I_C to change more than the change in I_{CBO}, it would be desirable to have $S = 1$. The formula for S is

$$S = \frac{R_E + R_B}{R_E + R_B/(h_{FE} + 1)}$$

where $h_{FE} = I_C/I_B$ and can be obtained from a manufacturer's data sheet.

For the circuit shown in Fig. 3-7,

$$S = \frac{R_E + (R_a R_b)/(R_a + R_b)}{R_E + [1/(h_{FE} + 1)]\, R_a R_b/(R_a + R_b)}$$

As previously mentioned, the value of S under ideal conditions is 1. From this it follows that S under worst-case conditions is equal to $1 + h_{FE}$. An S of 10 to 20 is very good, S of 5 is excellent, and an S of 1.1 is outstanding. Taking h_{FE} as 100 is often a very realistic choice.

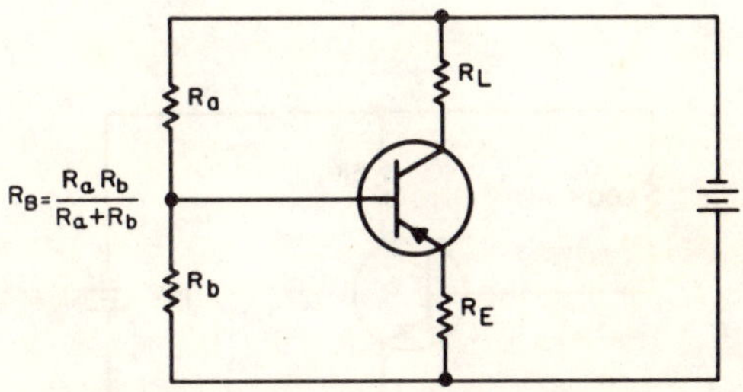

Fig. 3-7. Conventional biasing.

Now consider the simple circuit of Fig. 3-8. If $h_{FE} = 100$, $R_B = R_a$, since R_b is not in the circuit. Also $R_E = 0$. Under these conditions,

$$S = \frac{0 + R_B}{0 + R_B/(1 + h_{FE})} = 1 + h_{FE} \approx 100$$

From a stability point of view, the circuit of Fig. 3-8 is very unsatisfactory and should not be used unless the ambient temperature is regulated and the transistor is not working near its maximum values.

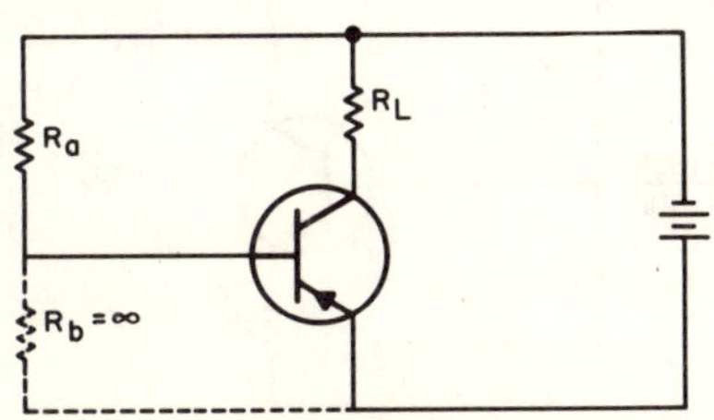

Fig. 3-8. Simple bias circuit with $R_b = \infty$.

The use of *emitter* bias is shown in Fig. 3-9. Assuming $h_{FE} = 100$,

$$S = \frac{1K + 600K}{1K + 600K/100} \approx \frac{600K}{7K} \approx 85$$

This represents little improvement over the circuit of Fig. 3-8. Thus, emitter bias, by itself, should be used with great care.

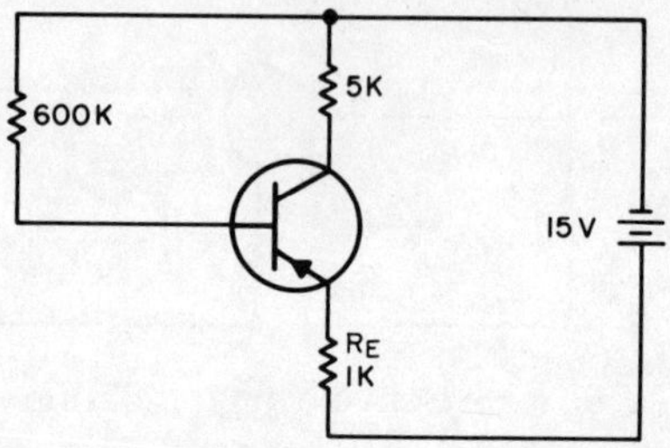

Fig. 3-9. Biasing with emitter feedback.

The arrangement of Fig. 3-10 is the most popular form of biasing and gives a stability factor that is satisfactory over a fairly wide range of temperatures. Using the values indicated,

$$S \approx \frac{1K + 9K}{1K + (9K/100)} = \frac{10K}{1.09K} \approx 9$$

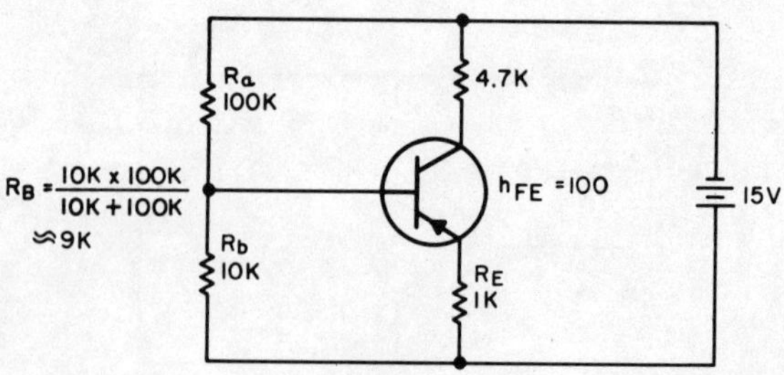

Fig. 3-10. A very popular form of biasing.

The circuit of Fig. 3-11 uses a form of biasing that is popular for some portable radios. It offers fair stability with minimum battery drain, and provides a certain amount of current feedback, which tends to stabilize the circuit gain.

Using the values shown in the illustration,

$$S = \frac{R_E + R_B}{R_E + R_B/(1 + h_{FE})} \approx \frac{6K + 400K}{6K + (400K/100)} \approx 40$$

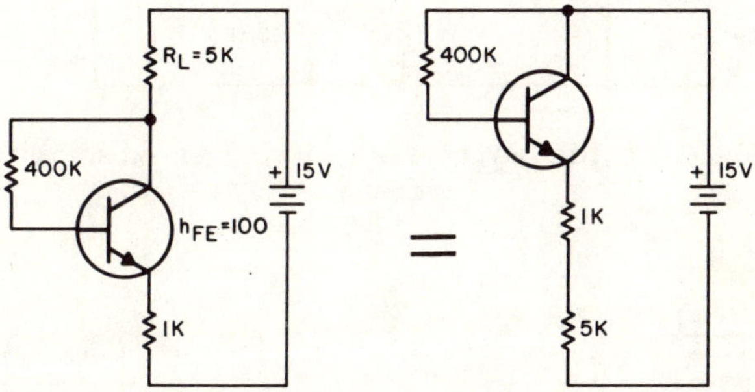

Fig. 3-11. Biasing with collector feedback.

The circuit shown in Fig. 3-12 will be considered in more depth later, but the stability factor is calculated here to show how a fairly sophisticated circuit is analyzed.

To find the value of S, the circuit is first redrawn as shown in Fig. 3-13A. The 10K resistor in the emitter lead is then *split* in terms of its relative contributions to base and collector biasing, respectively. This is indicated in Fig. 3-13B by the presence of resistances R_X and R_Y.

Now, since I_B is normally much smaller than I_C, let us assume that $I_C \approx I_E$ and that the bias divider current, termed I_{BD}, is approximately $10I_B$. Then, if $h_{FE} = 100$,

$$I_E \approx I_C$$
$$= 100I_B$$

and

$$I_C \approx 10I_{BD}$$

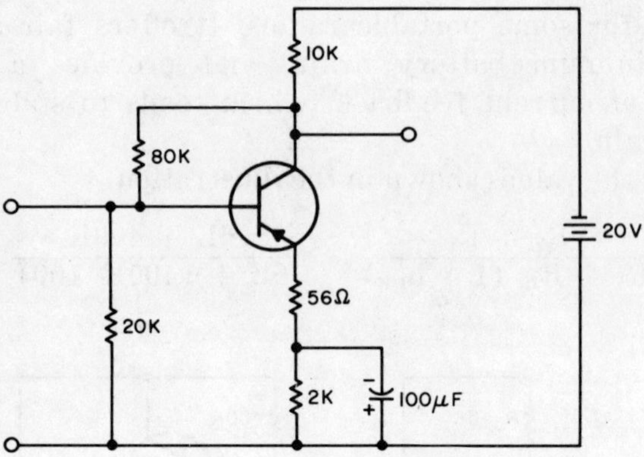

Fig. 3-12. Bias circuit for achieving a good stability factor.

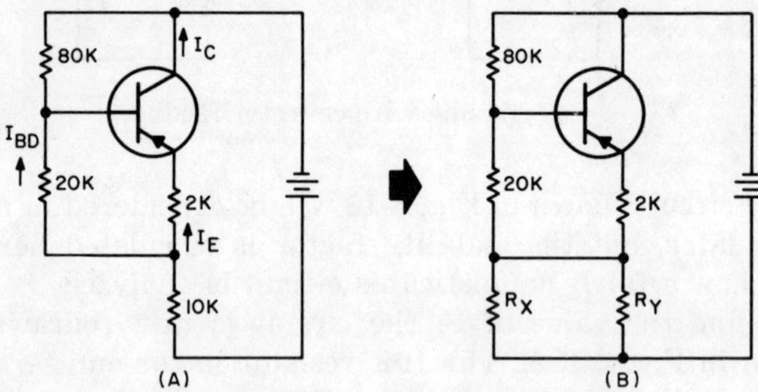

Fig. 3-13. Redrawn version of Fig. 3-12.

Therefore, $I_E = 10I_{BD}$

Thus, emitter current through the 10K resistor is 10 times the remaining current into the divider. One last requirement can now be met; viz, if the voltage drop across R_X is the same as across R_Y, there is no current through the short and it can be removed. From the above it can be seen that

$$R_X = 10\,R_Y \qquad \text{and} \qquad \frac{R_X R_Y}{R_X + R_Y} = 10K$$

or

$$\frac{(10R_Y)\,(R_Y)}{10R_Y + R_Y} = 10K$$

Therefore, $R_Y = 11K$ and $R_X = 110K$.

Having determined values for R_X and R_Y, we can redraw the circuit of Fig. 3-13B as shown in Fig. 3-14A. Finally, by combining the series-connected resistor with the base

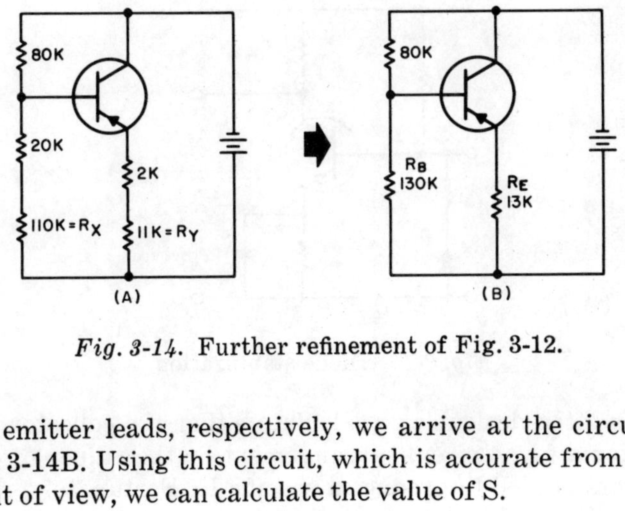

Fig. 3-14. Further refinement of Fig. 3-12.

and emitter leads, respectively, we arrive at the circuit of Fig. 3-14B. Using this circuit, which is accurate from a d-c point of view, we can calculate the value of S.

$$\begin{aligned} S &= \frac{R_E + R_B}{R_E + R_B/(1 + h_{FE})} \\ &= \frac{13K + 50K}{13K + 50K/100} \\ &= 4.6 \end{aligned}$$

(When measured in the laboratory, the S for this circuit was found to be 4.3.)

3.3 Biasing

A diode may also be used to stabilize the operation of a transistor circuit, as shown in Fig. 3-15. In this circuit, an increase in temperature causes a decrease in voltage across the diode. As a result, the forward bias on the transistor decreases and this compensates for the tendency of the transistor to conduct more than it did at the lower temperature. This particular configuration provides only partial compensation as the bias divider stick behaves somewhat like a constant-current source. Consequently, practical diode-compensation circuits are usually more complex.

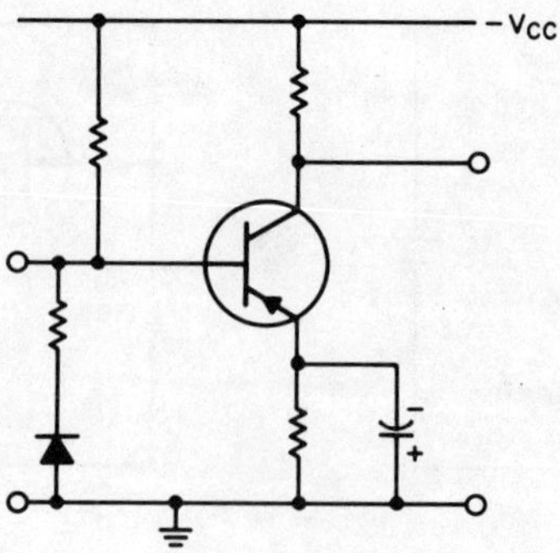

Fig. 3-15. Diode stabilization.

It is often desirable to calculate either the collector current for a given circuit arrangement or the values of bias resistors needed to produce a specific I_C. Fortunately, these calculations are easy to make.

Consider, for example, the circuit of Fig. 3-16. Suppose the transistor is a PNP germanium unit having an h_{FE} of 100. Notice that $h_{fe} = \Delta i_c / \Delta i_b$ and is an a-c quantity. Heretofore, we have used the d-c quantity $h_{FE} = I_C/I_B$. Now, suppose we have to determine the values of R_B for collector currents of 1 and 2 mA. First, let us determine R_B when $I_C = 1$ mA.

The value of V_{BE} for a germanium unit having $I_C = 1$ mA

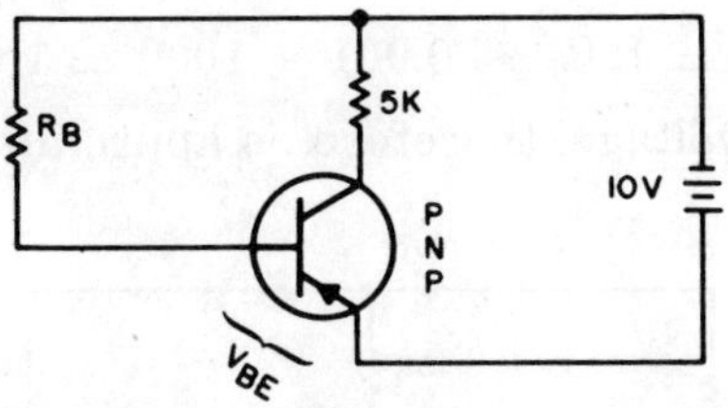

Fig. 3-16. Circuit used in determining value of R_b.

is approximately 0.2 volt, and, for practical purposes, is disregarded. Under these conditions, the voltage drop across R_B is 10 volts. For simplicity, this voltage drop is here indicated by the letter symbol V_B.

The solution is now a simple exercise in Ohm's law and noting that the a-c current gain h_{fe} is approximately equal to h_{FE}

$$R_B \approx \frac{V_B}{I_C/h_{fe}} \approx \frac{10}{0.001/100} \approx \frac{10 \times 100}{0.001} \approx 1\text{M}\,\Omega$$

When $I_C = 2$ mA,

$$R_B \approx \frac{10 \times 100}{0.002} \approx 500\text{K}$$

As another example of how to calculate the values of bias resistors for a specified I_C, refer to Fig. 3-17. Assume a germanium transistor with $h_{fe} \approx 100$ and $I_C \approx I_E$. Our problem is to determine the values of resistors R_a and R_b with $I_C = 1$ mA. We will also calculate the current stability factor S.

Since $h_{fe} \approx 100$ and $h_{fe} \approx h_{FE}$,

$$I_B = \frac{I_C}{h_{FE}} \approx \frac{0.001}{100} \approx 10\ \mu\text{A}$$

The emitter voltage is

$$V_E = I_E R_E \approx 0.001 \times 1000 = 1 \text{ volt}$$

and the base voltage, therefore, is approximately 1.2 volts.

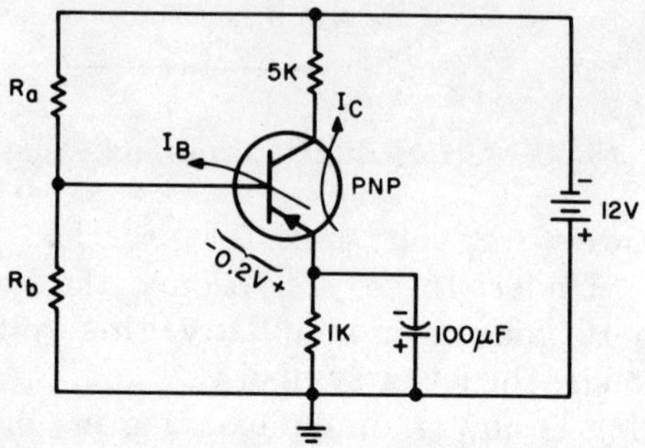

Fig. 3-17. Circuit for calculating bias resistor with a specified I_C.

If, for ease of calculation, the voltage divider bias stick current drain is made $10\,I_B$, the loading effect of I_B can be neglected and

$$R_b = \frac{V_B}{I_{BD}} \approx \frac{1.2}{100 \times 10^{-6}} = 12\text{K}$$

Similarly, $$R_a \approx \frac{10.8}{100 \times 10^{-6}} = 100\text{K}$$

Finally, $$S = \frac{R_E + R_B}{R_E + R_B/(1 + h_{FE})}$$

$$= \frac{1\text{K} + 11\text{K}}{1\text{K} + 11\text{K}/100} \approx 11$$

If I_E is the desired quantity, as it often is when A_e, the voltage gain, and R_i, the input impedance, are involved, the problem can be solved in reverse. That is, neglecting base-current loading effects, find V_B. Next, subtract the 0.2 V for Ge or 0.6 V for Si and find V_E. I_E, which is approximately the same as I_C, can be found directly by dividing V_E by the resistance in the emitter circuit.

CHAPTER 4

Amplifier Action

4.1 Why a Transistor Can Amplify

For purposes of explanation, an NPN transistor is considered. To understand how the transistor amplifies, there are three factors to keep in mind:

First, a carrier leaving the emitter enters a very narrow base. Because the base is so narrow, the probability of carrier recombination in the base is greatly reduced. Consider, for example, an electron leaving the emitter of an NPN transistor. When the electron enters the P-material base, it becomes a minority carrier and the probability that it will fall into a hole is quite high; however, the base is so narrow that most of the electrons find they are through the base region before they have had a chance to recombine.

Second, in an NPN device the percentage of electrons in the emitter is much greater than the percentage of holes in the base. This factor, coupled with the thickness of the base, makes it even easier for an electron to cross the base region without recombining.

Third, there is a strong accelerating field coming out of the collector.

Because of these three conditions (i.e., narrow base, relatively low doping in the base, high collector field), 95 to 99 percent of the electrons leaving the emitter of an NPN transistor will enter the collector. The 1 to 5 percent that recombine in the base become part of the base current. The rest of the base current is made up of carriers that are not swept out of the base by the field coming from the collector.

We can further clarify transistor action by taking the process step by step. Assume the emitter-base junction of an NPN device is forward biased and the collector lead is momentarily opened. The emitter-base potential establishes a certain emitter-base current. When the open collector is reconnected, the combination of a narrow base, and the positive field coming out of the collector attracts 95 to 99 percent of the electrons to the collector. If nothing else happened, the base current would fall to about 2 percent of its original value. However, the emitter-base junction is a diode and must pass the current dictated by its bias conditions. Instead of the base current falling, it stays constant and, due to the previously described action, the collector current rises to a much higher value. The ratio of this d-c collector current to the d-c base current is called h_{FE} or β_{dc}. A typical value of h_{FE} is 100 for a small-signal, low-frequency transistor although it may vary from near 30 to over 200. Since variations in base current show up as similar variations in collector current, and since the collector current is always larger than the base current, the device is said to have *gain* (amplification). If an input signal is applied to the base-emitter junction, a larger output signal can be taken between the collector and emitter. This configuration is called a *common-emitter* (CE) circuit and is characterized by having both voltage and current gains.

It should be noted that if the base of an NPN transistor is *not* connected to a potential, the base region is wide enough to prevent most electrons from flowing directly from the emitter to the collector. The emitter-base circuit is needed to get the electrons into the base where they can come under the influence of the collector field. That is, the emitter-base bias starts the electrons on their journey and the collector field finishes the job.

If PNP devices are used instead of NPN devices, the logic is the same but the carriers leaving the emitter are holes instead of electrons.

4.2 Graphic Analysis

To see what is happening in a transistor stage, the most important area to examine is the *base-emitter junction.* If the base-emitter diode is forward biased, the transistor is conducting (on). If it is heavily forward biased, it may be saturated (essentially a closed switch). If the base-emitter junction is reverse-biased, the transistor is non-conducting (cut off).

If a germanium transistor is lightly forward biased, e.g., 1-mA collector current and 0.2-V forward bias on the base-emitter junction, the transistor stage is acting as an amplifier. Consider the common (grounded) emitter circuit shown in Fig. 4-1.

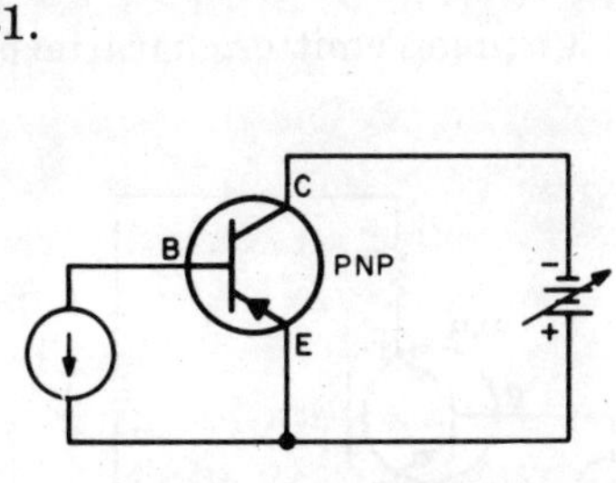

Fig. 4-1. Common-emitter operation.

If collector current I_C is plotted as a function of collector-to-emitter voltage (V_{CE}), a set of curves similar to those shown in Fig. 4-2 is obtained. These curves are obtained by point-to-point plotting or by using a curve tracer. Note that $\Delta I_C >> \Delta I_B$. Thus, the a-c current gain is greater than unity.

Now, to consider the CE configuration as an amplifier, refer to Fig. 4-3. To simplify the illustration, all components used for stabilization of the operating point are omitted. Notice that

$$v_{ce} = V_{CC} - i_c R_L$$

or

$$i_c = \frac{-v_{ce}}{R_L} + \frac{V_{CC}}{R_L}$$

$$= -\frac{1}{R_L} v_{ce} + \frac{V_{CC}}{R_L}$$

Of course, i_c and v_{ce} are variables and V_{CC} and R_L are constants.

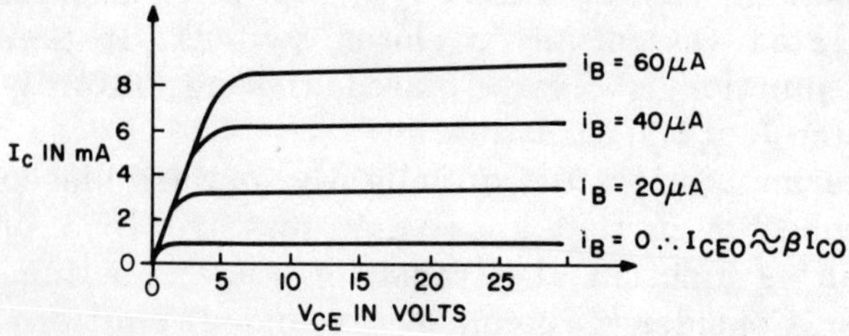

Fig. 4-2. Common-emitter characteristics.

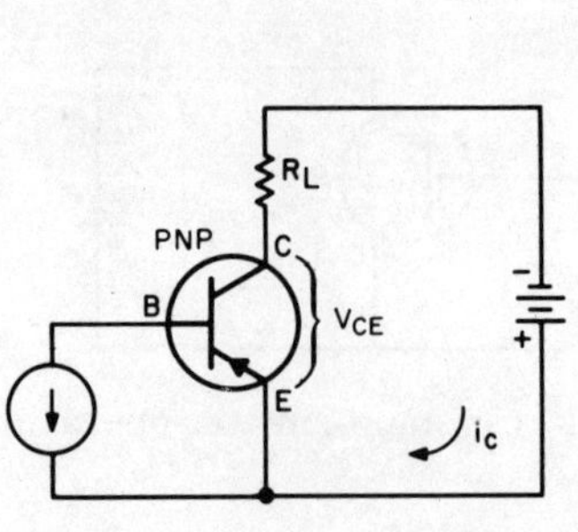

Fig. 4-3. A simple CE arrangement.

Since there are no squared or cubed terms in the final equation noted above, it is the equation of a straight line. Two points define this line; viz., V_{CC}/R_L and V_{CC}. Refer to Fig. 4-4.

We now have two equations for i_c: the straight-line equation illustrated is Fig. 4-4, and an equation characterized by the curves of Fig. 4-2, with i_c a function of v_{ce}

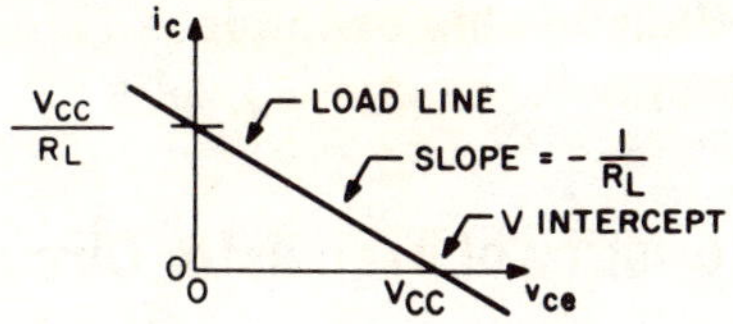

Fig. 4-4. Constructing a load line.

and I_B. The two equations can be solved graphically, as shown in Fig. 4-5.

With the load line directly on the collector family of curves, note in Fig. 4-6 how the operating point moves as i_b is varied. The solution point (operating point) moves along the load line and satisfies both equations as i_b is

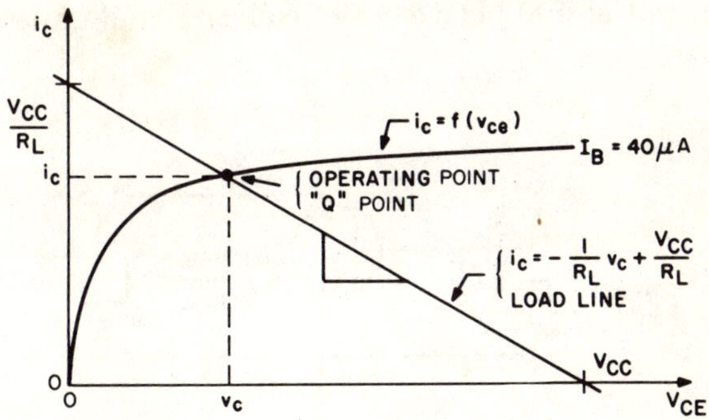

Fig. 4-5. The operating (Q) point.

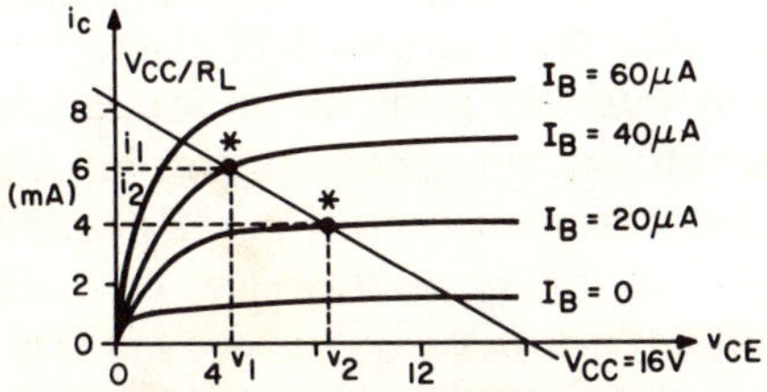

Fig. 4-6. Load line superimposed on characteristics.

varied. Notice that, in this example, a change of 20 μA in base current changes v_{ce} by about 4 volts.

4.3 Characteristics of Transistor Circuits

There are three transistor configurations used in practice; namely, the common-emitter, common-base, and common-collector (emitter-follower) circuits.

The most commonly used arrangement is the *common-emitter* circuit shown in Fig. 4-7. This is the only circuit configuration with both voltage and current gains greater than unity. Accordingly, it has the best available power gain. It is also the only circuit configuration that provides phase inversion; that is, a positive-going input signal applied between the base and emitter appears as a negative-going output signal between the collector and emitter.

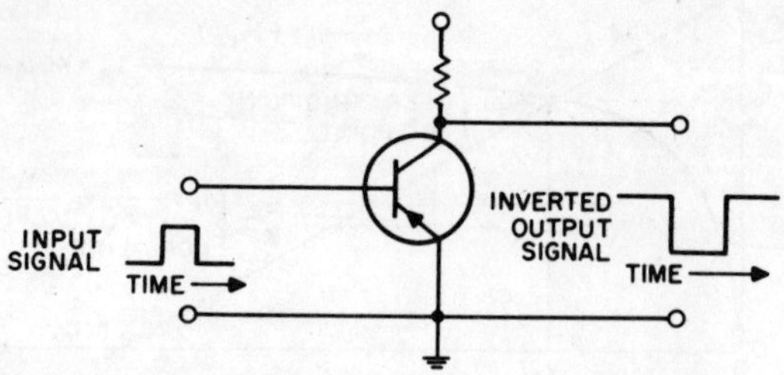

Fig. 4-7. The CE amplifier as an inverter.

Probably the next most widely used circuit configuration is the *common-collector* (emitter-follower) circuit, Fig. 4-8, which has good current gain, high input impedance, low output impedance, and no phase inversion. The most common application for this circuit is impedance matching; for example, the output impedance can be made low enough to drive a speaker coil directly, or two emitter followers in series can produce an input impedance of several megohms.

The *common-base* circuit configuration, Fig. 4-9, has low

input impedance, high output impedance, very good voltage gain, and current gain less than unity. Like the common-collector, it does not provide phase inversion.

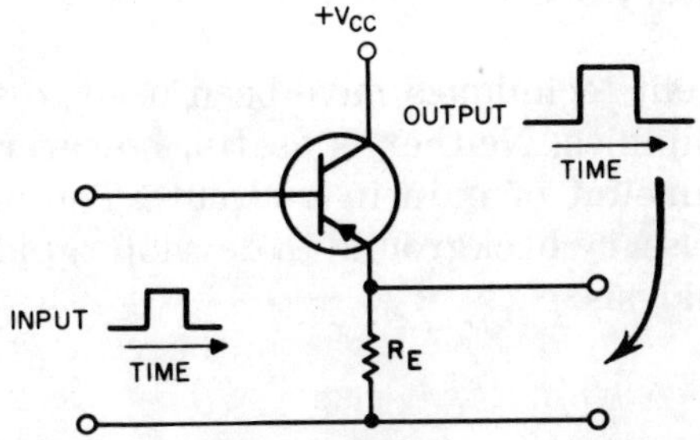

Fig. 4-8. The common-collector (emitter follower) circuit.

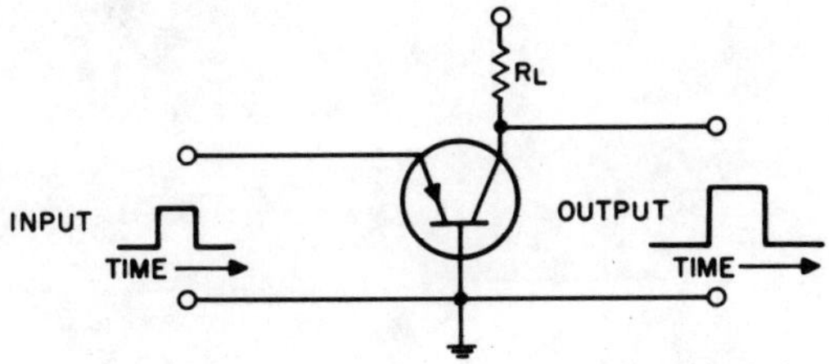

Fig. 4-9. The common-base circuit.

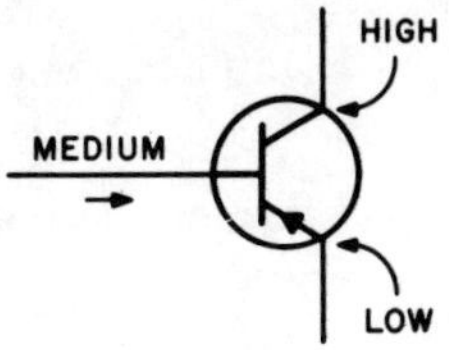

Fig. 4-10. General impedance characteristics.

The resistors and capacitors associated with a transistor circuit have a direct bearing on the various impedance levels around the transistor. In general, however, the impedance levels are as is shown in Fig. 4-10.

All circuit characteristics are summarized in Appendix 1, which shows rule-of-thumb formulas.

4.4 Summary

Two different techniques have been used to show how a transistor amplifies. Neither is useful, however, to quickly predict the amount of gain in a circuit. The next chapter gives the necessary background to develop rapid techniques for circuit analysis.

CHAPTER 5

The H-Parameter Equivalent Circuit

To accurately predict the behavior of a transistorized circuit it is easiest to simulate the transistor with a model consisting of voltage and current sources and resistors. If the model is an accurate representation of the transistor at the frequencies of interest, then it should be a fairly simple matter to predict the voltage gain (A_e), the current gain (A_i), the input impedance (R_i), and the output impedance (R_o). The hybrid (h) parameter model does this quite well at medium and low frequencies.

5.1 The H Parameters

Any amplifier, whether a single stage or a complete circuit, can be considered as a *black box* that has two input terminals and two output terminals, as shown in Fig. 5-1. The quantities, I_1, I_2, V_1, and V_2 are measured at the input

Fig. 5-1. The black-box concept.

and output terminals, with I_1 and I_2 both considered positive (flowing inward) so that specification of input or output does not alter the analysis. Knowing the values of these circuit quantities, we can specify the performance of the circuit.

In the analysis, two quantities are taken as independent variables. The choice of which two quantities are considered dependent variables and which two are considered independent variables is arbitrary, but it is this choice that determines the form of the equivalent circuit.

Suppose the black box of Fig. 5-1 contains a transistor amplifier in the *common-base* configuration, and we select emitter current i_e and collector-to-base voltage v_{cb} as the independent variables. We can write the general equation:

$$v_{eb} = f\,(i_e, v_{cb})$$

and

$$i_c = f\,(i_e, v_{cb})$$

To develop an *input circuit* equation from the general equation, we must first determine the effect of *each* independent variable on v_{eb} and then add the separate results. When the effect of i_e is considered, v_{cb} is held at zero, and vice versa.

For small variations of I_E and V_{CB} from quiescent values, we can write:

$$\Delta v_{eb} = \Delta i_e \left| \frac{\Delta v_{eb}}{\Delta i_e} \right|_{v_{cb}=0} + \Delta v_{cb} \left| \frac{\Delta v_{eb}}{\Delta v_{cb}} \right|_{i_e=0}$$

To develop an output circuit equation from the general equation, we proceed in the same manner and obtain

$$\Delta i_c = \Delta i_e \left| \frac{\Delta i_c}{\Delta i_e} \right|_{v_{cb}=0} + \Delta v_{cb} \left| \frac{\Delta i_c}{\Delta v_{cb}} \right|_{i_e=0}$$

The first coefficient of the input equation is called the *input resistance* and is measured in *ohms* because it represents the ratio of a voltage to a current. The h parameter symbol for this quantity is h_{ib}, where the subscripts i and b represent input and common base, respectively. Using numerical subscripts, h_{ib}, is written as h_{11b}.

The second coefficient of the input equation is *dimensionless* and is called the *reverse (feedback) voltage ratio.* The h-parameter symbol is h_{rb}, or h_{12b}.

The first coefficient of the output equation is *dimensionless* and is called the *forward current transfer ratio.* The h-parameter symbol is h_{fb}, or h_{21b}.

Finally, the second coefficient of the output equation is called the *output conductance* because it represents the reciprocal of resistance and is measured in *mhos.* The h-parameter symbol is h_{ob}, or h_{22b}.

We now have four h parameters for the common-base configurations. Using the numerical forms of these parameters, we can rewrite the input and output equations as:

$$v_{eb} = h_{11b}i_e + h_{12b}v_{cb}$$
$$= h_{ib}i_e + h_{rb}v_{cb}$$

and

$$i_c = h_{21b}i_e + h_{22b}v_{cb}$$
$$= h_{fb}i_e + h_{ob}v_{cb}$$

The input equation suggests an equivalent circuit of a resistance in series with a voltage generator, and the output equation suggests an equivalent circuit of a current generator in parallel with a conductance. If we combine these two, we obtain the equivalent circuit shown in Fig. 5-2.

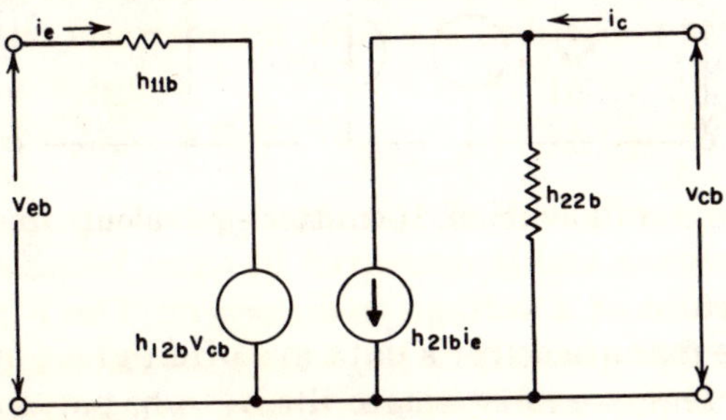

Fig. 5-2. The common-base equivalent circuit.

In the common-emitter circuit, i_b and v_{ce} are the independent variables, and the general equations are

$$i_c = f(i_b, v_{ce})$$
$$v_{be} = f(i_b, v_{ce})$$

and, by following the methods previously detailed, the following h-parameter equations are obtained:

$$v_{be} = h_{11e}i_b + h_{12e}v_{ce}$$
$$= h_{ie}i_b + h_{re}v_{ce}$$

and

$$i_c = h_{21e}i_b + h_{22e}v_{ce}$$
$$= h_{fe}i_b + h_{oe}v_{ce}$$

where the second subscript e indicates that we are dealing with the common-emitter configuration.

The above equations suggest the equivalent circuit of Fig. 5-3. Notice that this is the same form as that of the common-base amplifier but with different parameter values. Thus, the gain and impedance equations to be derived apply equally well to either configuration if the appropriate h parameter is used.

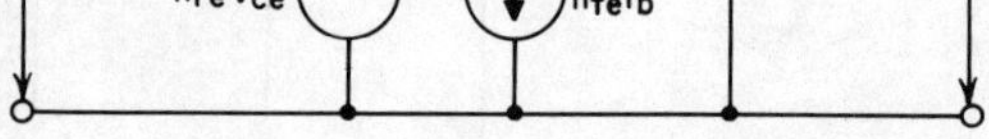

Fig. 5-3. The common-emitter equivalent circuit.

Because manufacturer's data are often given in terms of h_b parameters, we must relate the h_b to h_e parameters. For all *practical* purposes, the following relationships suffice:

$$h_{11e} = \frac{h_{11b}}{1 + h_{21b}} \quad \text{or} \quad h_{ie} = \frac{h_{ib}}{1 + h_{fb}} \approx h_{ib}h_{fe}$$

$$h_{12e} = \frac{\Delta h_{12b}}{1 + h_{21b}} \quad \text{or} \quad h_{re} = \frac{\Delta h_{rb}}{1 + h_{fb}} \approx \frac{1}{1 + h_{fb}}$$

$$h_{21e} = -\frac{h_{21b}}{1 + h_{21b}} \quad \text{or} \quad h_{fe} = -\frac{h_{fb}}{1 + h_{fb}}$$

$$h_{22e} = \frac{h_{22b}}{1 + h_{21b}} \quad \text{or} \quad h_{oe} = \frac{h_{ob}}{1 + h_{fb}} \approx h_{ob}h_{fe}$$

where $\Delta = h_{ie}h_{oe} - h_{rb}h_{fb}$ and h_{fb} is a negative number.

Notice that the quantity $-h_{21b}$ in the third equation above is equal to the current gain of the common-base configuration, often designated by the letter symbol α. By substitution,

$$h_{21e} = \frac{\alpha}{1 - \alpha} = \beta$$

or

$$h_{fe} = \beta$$

where β is the a-c current gain of the common-emitter configuration.

If the data are given in h_e parameters and conversions to h_b parameters are required,

$$h_{ib} = \frac{h_{ie}}{1 + h_{fe}} \approx \frac{h_{ie}}{h_{fe}} \qquad h_{fb} = \frac{-h_{fe}}{1 + h_{fe}} \approx \frac{1}{h_{fe}} - 1$$

$$h_{rb} = \frac{\Delta - h_{re}}{1 + h_{fe}} \qquad h_{ob} = \frac{h_{oe}}{1 + h_{fe}} \approx \frac{h_{oe}}{h_{fe}}$$

where $\Delta = h_{ie}h_{oe} - h_{re}h_{fe}$.

5.2 Summary

Probably the most useful conversion relationships given are $h_{ie} \approx h_{ib}h_{fe}$ and $h_{fe} \approx 1/(1 + h_{fb})$, which form the foundation of the simplified analysis developed in the next chapter.

CHAPTER 6

Simplified Circuit Analysis

In addition to the conventional h-parameter techniques for predicting circuit performance, there is a set of simplified expressions that can be used for the same purpose. These simplified expressions are just logical extensions of the h-parameter notation but are reduced to only the essential terms. They are summarized as rule-of-thumb expressions in Appendix 1.

To show the origin of this simplified technique, several of the more important expressions will be derived from the h-parameter model.

6.1 Deriving Simplified Expressions

A typical transistor amplifier in the common-emitter configuration is shown in Fig. 6-1. The equivalent circuit of this amplifier, when operating over the midfrequency range, is shown in Fig. 6-2. In the equivalent circuit, e_g represents an a-c signal source and R_g represents the internal resistance of the source; i.e., Thévenin's equivalent circuit.

To solve for the current gain, A_i, notice that in the right-hand portion of the equivalent circuit:

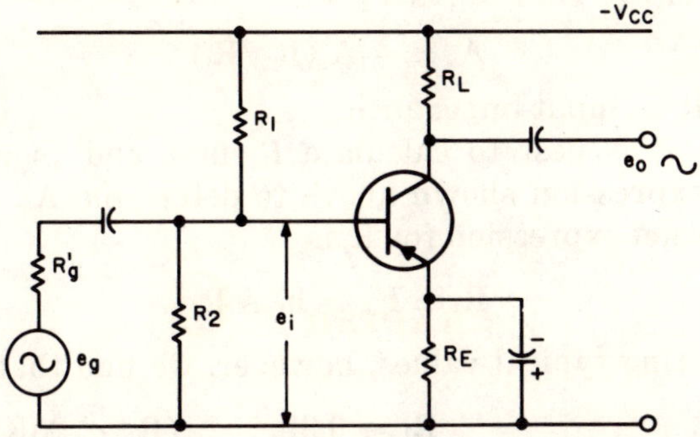

Fig. 6-1. A typical common-emitter amplifier circuit.

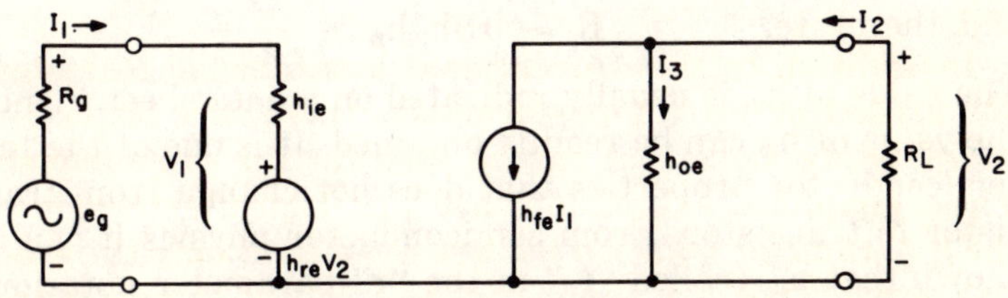

Fig. 6-2. The equivalent circuit of Fig. 6-1 at midfrequencies.

$$-I_2R_L = V_2$$

and $$V_2h_{oe} = I_3$$

However, $$I_2 = I_3 + h_{fe}I_1$$

Therefore, $$I_2 = V_2h_{oe} + h_{fe}I_1$$

and $$I_2 = -\,I_2R_Lh_{oe} + h_{fe}I_1$$

so $$I_2(1 + h_{oe}R_L) = h_{fe}I_1$$

and thus $$A_i = \frac{I_2}{I_1} = \frac{h_{fe}}{1 + h_{oe}R_L}$$

Typically, if $R_L < 20K$,

$$A_i \approx 0.8h_{fe}$$

The voltage gain, A_e, is V_2/V_1.

so $$A_e = -A_i(R_L/R_i)$$

where R_i = input impedance.

Thus, it is best to calculate R_i first and then use the simple expression shown above to determine A_e.

The exact expression for R_i is

$$R_i = h_{ie} - h_{re}A_iR_L$$

By inserting typical values, however, we find that:

$$R_i \approx 0.8h_{ie} \qquad (R_L < 20K)$$

Also, from the previous chapter,

$$h_{ie} \approx h_{ib}h_{fe}$$

and, therefore, $$R_i \approx 0.8h_{ib}h_{fe}$$

The value of h_{ib} is usually indicated on a data sheet. If not, the value of h_{ib} can be readily obtained. It is one of the few semiconductor properties that does not change from transistor to transistor. From semiconductor physics it can be shown that h_{ib} (called "r_e" in the "r"-parameter notation) has a theoretical value of 26 ohms/$|I_E(mA)|$. However, due to lead resistances and connections to the semiconductor material, a more realistic value is 30 ohms/$|I_E(mA)|$. That is, if the emitter current is 1 mA, h_{ib} behaves as an *unbypassed* emitter feedback resistor of 30 ohms. At 2 mA, h_{ib} would be similar to a 15-ohm resistor; and at 0.5-mA, it would be 60 ohms. These values are accurate for both germanium and silicon planar transistors.

Using the above relationships,

$$A_e \approx \frac{-0.8h_{fe}R_L}{0.8h_{fe}h_{ib}}$$

$$\approx -\frac{R_L}{h_{ib}}$$

In practice, it is found that the above approximation best be modified to:

$$A_e = \frac{-0.9R_L}{h_{ib}}$$

and that the accuracy of the calculation increases as h_{ib} increases, i.e., as emitter current decreases. The reason for the modification factor of 0.9 can be seen in Fig. 6-3, which also gives correction factors when $R_L > 20K$.

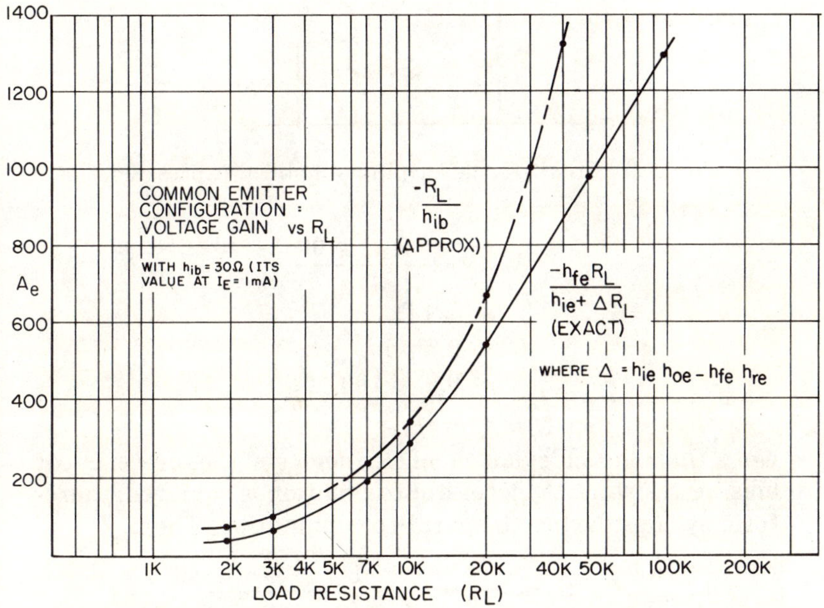

Fig. 6-3. Voltage gain (A_e) as a function of load resistance.

6.2 Sample Calculations

To illustrate the application of the simplified expressions, let us first consider the single stage, common-emitter amplifier shown in Fig. 6-4. The transistor is considered to be *any* small-signal, medium-frequency device; e.g., the 2N1303. If $I_E \approx 1$ mA, $h_{ib} \approx 30$ ohms. We can now calculate the values of A_i, A_e, R_i, and R_o almost by inspection.

$$A_e \approx \frac{-0.9R_L}{h_{ib}}$$

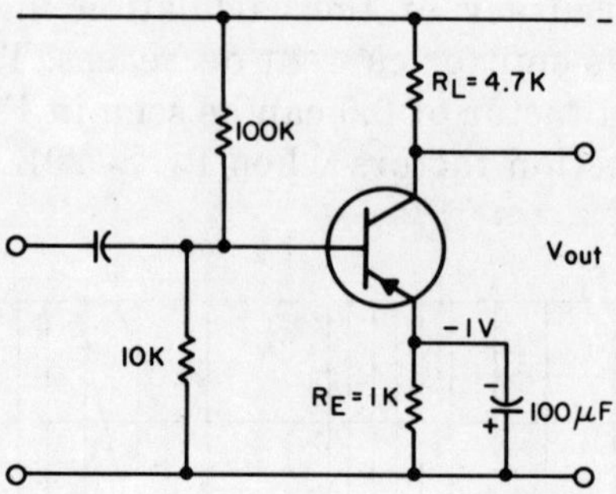

Fig. 6-4. Circuit for developing simplified expressions.

$$A_e \approx \frac{-0.9 \times 4700}{30}$$
$$\approx -141$$
$$R_i \approx 0.8h_{fe}h_{ib}$$
$$\approx 0.8 \times 100 \times 30$$
$$\approx 2.4K$$

Note that the calculation of R_i neglects the presence of biasing resistors. A closer approximation is achieved, therefore, by considering the parallel combination. Then,

$$R_i \approx \frac{1}{\frac{1}{2.4K} + \frac{1}{10K} + \frac{1}{100K}}$$
$$\approx 2K$$
$$A_i \approx 0.8h_{fe}$$
$$\approx 0.8 \times 100$$
$$\approx 80$$

and $$R_o \approx R_L$$
$$\approx 4.7K$$

Multistage amplifiers are handled just like single-stage amplifiers. They are broken down into single stages and solved individually. The only other item that must be considered is the effect of the interstage coupling.

A typical three-stage amplifier is shown in Fig. 6-5. At

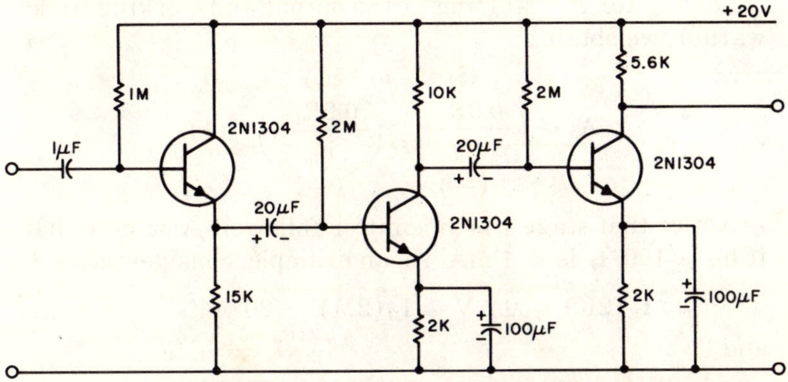

Fig. 6-5. A typical three-stage amplifier.

midfrequencies the reactance of the capacitors is negligible so the circuit can be redrawn as shown in Fig. 6-6. Using this circuit, we can calculate A_e, A_i, R_i, R_o, and the power gain, A_p.

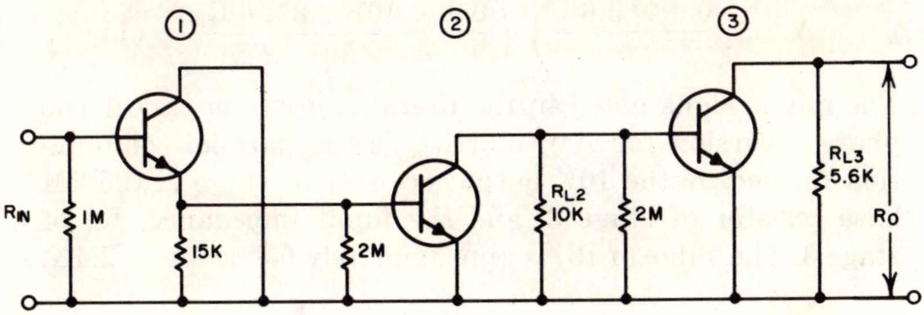

Fig. 6-6. Simplified version of the three-stage amplifier.

For the 2N1304, $h_{fe} \approx 100$, and $h_{ib} \approx 30$ ohms when $I_E = 1$ mA. Using the relationship, $h_{ie} \approx h_{ib}h_{fe}$, $h_{ie} = 3K$.

It should be noted that typical data sheet values are $h_{ib} \approx 28$ ohms when $I_E = 1$ mA. Also, $h_{fe} \approx 120$ and $h_{ie} \approx 3.3K$. These values compare favorably with the typical values noted above.

Solving for A_e (starting at the output and working backwards), we obtain:

$$A_e \approx \underset{(-)}{\overset{③}{\left(\frac{0.9R_{L3}}{h_{ib}}\right)}} \underset{(-)}{\overset{②}{\left(\frac{0.9R_{L2}}{h_{ib}}\right)}} \underset{(+)}{\overset{①}{(1)}}$$

Notice that stage 1 is an emitter follower. Also note that if $h_{fe} \approx 100$, I_E is ≈ 1 mA. As an example, consider stage 3.

$$I_E(2K) + 0.2\text{ V} + I_B(2M) = 20 \text{ volts}$$

and

$$I_E \approx I_C = h_{FE}\, I_B$$

$$I_E(2K) + 0.2\text{ V} + (I_E/100)\,2M = 20 \text{ volts}$$

$$I_E(2K) + I_E(20K) \approx 20 \text{ volts}$$

$$I_E \approx 20/22 \text{ mA}$$

$$= 0.9 \text{ mA}$$

Inserting numerical values into the equation for A_e, we obtain:

$$A_e = \left(\frac{0.9 \times 5.6K}{30}\right)\left(\frac{0.9 \times 10K \parallel 2M \parallel R_{i3}}{30}\right)(1)$$

The minus signs noted in the literal expression cancel (no phase inversion for A_e), and R_{L2} is the parallel combination formed by the 10K output resistor of stage 2, the 2M base resistor of stage 3, and the input impedance, R_i, of stage 3. The value of R_{i3} is approximately $0.8 \times 3K \approx 2.4K$.

$$\text{Now, } A_e \approx \left(\frac{\cancel{0.9} \times 5.6\cancel{K}}{\cancel{30}}\right)\left(\frac{0.9 \times \overset{2K}{\cancel{10K} \parallel \cancel{2M} \parallel \cancel{2.4K}}}{\cancel{30}}\right)(1)$$

$$\approx 5.6 \times 0.9 \times 2K \approx 5.04 \times 2K = 1 \times 10^4 = 10{,}000$$

Where the symbol $\parallel$ indicates parallel quantities,

$$R_{in} \approx 1M \parallel \overset{80}{\cancel{0.8} \times \cancel{100}}\,(15K \parallel \overset{2.4K}{\cancel{2M} \parallel \cancel{0.8} \times \cancel{3K}})$$

$$\approx 1M \parallel 80 \times 2K \approx 1M \parallel 160K \approx 140K$$

$$A_i \approx -A_e \frac{R_i}{R_L} \approx -10^4 \times \frac{140K}{5.6K} = -2.5 \times 10^5$$

Finally,

$$R_o \approx R_{L3} \approx 5.6K$$

and, $A_p \approx 10 \log (1 \times 10^4)(2.5 \times 10^5)$

$$= 90 \text{ dB} + 10 \log 2.5$$

$$= 90 \text{ dB} + 4 \text{ dB} \approx 94 \text{ dB}$$

For the most part, h_{fe} has been neglected in the foregoing discussion of voltage gain. Notice that in the formula $A_e \approx 0.9R_L/h_{ib}$, it doesn't even appear. The real significance of h_{fe} is in its influence on the transistor input impedance, which, in a multistage amplifier, is in parallel with the previous stage load resistor. As an example, if a grounded-emitter transistor has a low h_{fe}, it will have a proportionally low input impedance, which, in turn, will lower the voltage gain of the previous stage. This effect can be summarized as follows:

1. To increase the voltage gain of a single-stage amplifier, the d-c collector current should be increased (reduces h_{ib}).
2. In multistage amplifiers, the transistor current gain h_{fe} indirectly affects voltage gain. This term h_{fe} affects the input impedance, which appears in shunt across the previous stage load resistor.
3. To increase the voltage gain of a multistage amplifier, only the d-c current in the first gain stage should be increased. In all other stages, the improvement is neutralized by the loading effect on the previous stage.

A circuit with several unusual properties is shown in Fig. 6-7. It has voltage feedback (the 56-ohm emitter resistor), current feedback (the 80K collector-to-base resistor), and a biasing arrangement that gives an excellent stability factor. The stability factor, S, was calculated in a previous section and was found to be 4.6.

Quite often it is either helpful or necessary to calculate A_e, A_i, R_i, and R_o. Exact calculations for this circuit would be quite involved, but using the previously-developed simplified expressions, a considerable amount of information can be obtained rapidly. In the calculations that follow, use is made of the rule-of-thumb equations presented in Appendix 1.

Because voltage feedback is used in the circuit of Fig. 6-7, A_e is calculated by using the voltage feedback equation:

$$A_e \approx -\frac{R_L \parallel R_f}{h_{ib} + R_E}$$

where || means *in parallel.*

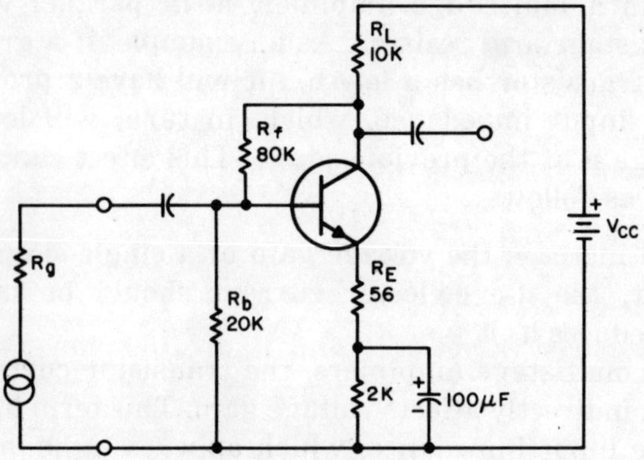

Fig. 6-7. Amplifier with emitter and collector feedback.

Substituting numerical values,

$$A_e \approx -8.9K/86$$
$$\approx -100$$

The minus sign indicates phase inversion by the single-stage, common-emitter amplifier.

The circuit of Fig. 6-7 also uses current feedback, so A_i is calculated by using the current feedback equation presented in Appendix 1.

$$A_i \approx R_f/R_L$$
$$\approx 80K/10K$$
$$\approx 8$$

It is interesting to note that the measured values of A_e and A_i are 100 and 7.1, respectively.

To calculate R_i, we notice that $R_f >> R_L$ and use the equation:

$$R_i = \frac{R_f(h_{ib} + R_E)}{R_f \parallel R_L}$$
$$\approx \frac{80{,}000 \times 86}{8900}$$
$$\approx 760 \text{ ohms}$$

In the calculation of R_o, we assume $R_g = 600$ ohms.

$$R_o \approx R_f(h_{ib}/R_g + 1/h_{fe}) \parallel R_L$$
$$\approx 80K(30/600 + 1/100) \parallel 10K$$
$$\approx 80K(1/20 + 1/100) \parallel 10K$$
$$\approx 80K \times 6/100 \parallel 10K$$
$$\approx 800 \times 6 \parallel 10K$$
$$\approx 4.8K \parallel 10K$$
$$\approx 3.25K$$

The measured values of R_i and R_o are 750 ohms and 3.5K ohms, respectively. R_i was measured with an R_g of 600 ohms.

6.3 The Emitter Follower

Since the common collector, which is more commonly called the emitter follower, has several very unique characteristics, it is worthwhile discussing in detail. For example, from pages 110 and 113 we can see there is no voltage phase inversion ($+A_e$) but there is 180° of current phase inversion ($-A_i$). The fact that there is no voltage inversion is quite useful as the very popular grounded emitter gives 180° of voltage phase inversion.

Three items of information are often wanted in emitter-follower circuits. These are the voltage gain, the input impedance, and the output impedance. Luckily, by using our simplified analysis each item is easily obtained.

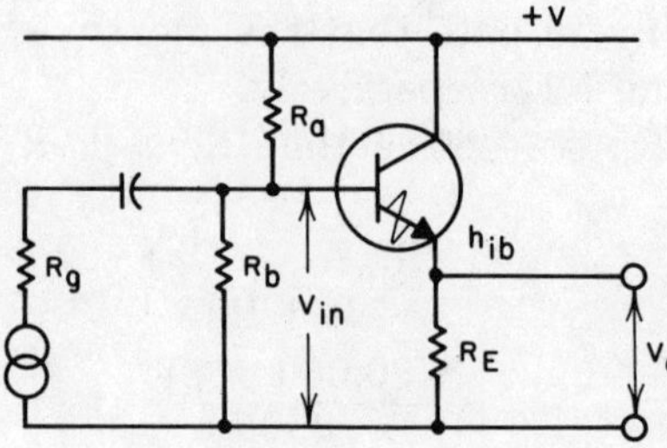

Fig. 6-8. Emitter-follower circuit.

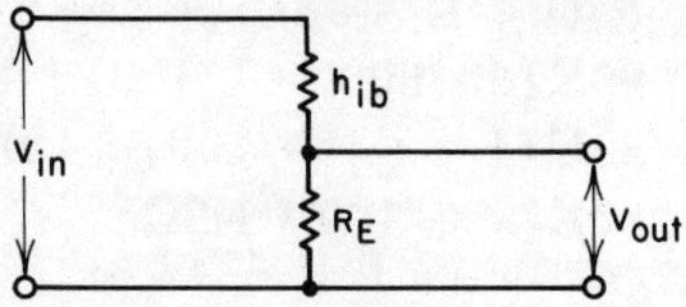

Fig. 6-9. Gain section of emitter follower.

First the voltage gain. Figure 6-8 shows a typical emitter-follower circuit. If we define the voltage gain to be V_o/V_{in}, the voltage from the emitter-to-ground divided by the voltage from base-to-ground, we can easily obtain the voltage gain expression. The input, V_{in}, appears inside the transistor just above the built-in feedback resistor. Then Fig. 6-8 reduces to Fig. 6-9. Now by inspection the approximate gain expression becomes:

$$A_e = +\frac{R_E}{R_E + h_{ib}}$$

Note that A_e for the emitter follower is always less than 1. A little bit of reduction of the exact voltage gain formula and, in the last step, eliminating those components which have a very small influence give the same result. Interestingly , this expression is one of the most accurate of all the approximate formulas.

What the above expression shows is that if R_E is low—a loudspeaker for example—then h_{ib} must be very small or you will lose considerable signal. This follows intuitively as you know you need a power transistor with considerable current to drive a loudspeaker. The high current keeps h_{ib} low and allows the signal to be coupled into the loudspeaker without much loss.

The expression for input resistance is the same as for the grounded emitter.

$$R_i \approx .8\, h_{fe}\, (R_E + h_{ib})$$

In most cases R_E is much greater than h_{ib} so this expression usually reduces to

$$R_i \approx .8\, h_{fe} R_E$$

Also when R_E is sufficiently large, the 0.8 is less critical and

$$R_i \approx h_{fe} R_E \quad \text{is a good approximation.}$$

The output resistance can also be found relatively easily. We have already shown that the emitter resistance can be moved into the base circuit by multiplying by h_{fe}. The same principle works the other way. That is, we can move a resistance from the base circuit to the emitter circuit by *dividing* by h_{fe}.

Consider the circuit of Fig. 6-10 which is a simplified version of Fig. 6-8. Moving R_g to the emitter side of the base by *dividing* by h_{fe} yields Fig. 6-11.

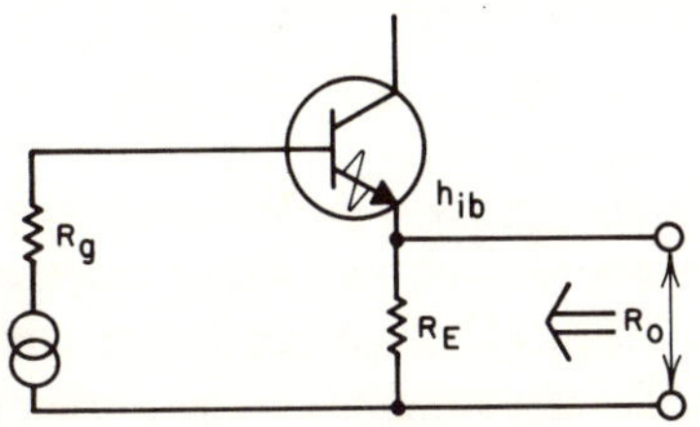

Fig. 6-10. Simplified emitter follower.

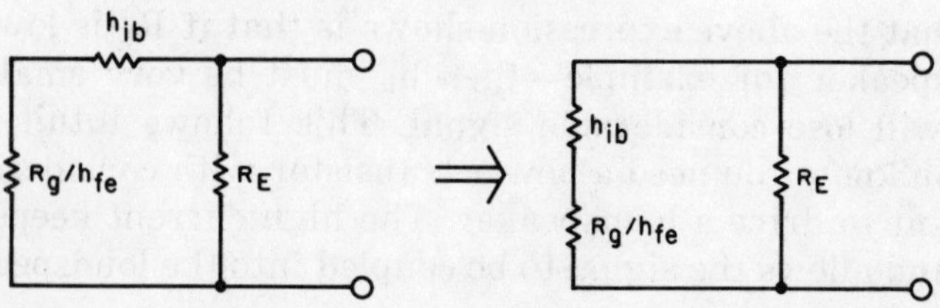

Fig. 6-11. Equivalent output of emitter follower.

Figure 6-11 shows several things. *First,* the output is generally low. For example, consider an emitter follower with $R_E = 10K$, $R_g = 600\Omega$, and $I_E = 2ma$. Then R_o is 10K in parallel with 15 ohms plus 6 ohms, giving an output of approximately 21 ohms. *Second,* the generator or driving source impedance greatly influences the emitter-follower output impedance. *Finally,* the output impedance can be modified relatively easily by changing the emitter current which, in turn, changes h_{ib}.

6.4 Summary

The calculations show the speed and acceptable accuracy obtained by using the rule-of-thumb equations given in Appendix 1. Since transistor values are rarely known to be closer than 20 percent (unless they are measured), the rule-of-thumb equations are adequate for all *practical* situations.

CHAPTER 7

Feedback

In the equation $A_e \approx -(0.9R_L/h_{ib})$, the term h_{ib} may be considered as a small resistor in series with the emitter load. See Fig. 7-1. This effective resistance produces a small amount of degenerative feedback. As I_E increases, h_{ib} decreases in a fairly constant manner. From this it follows that as the signal current varies, h_{ib} also varies, and

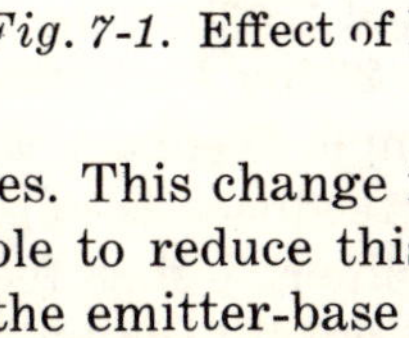

Fig. 7-1. Effect of h_{ib}.

voltage gain A_e varies. This change in A_e shows up as distortion. It is desirable to reduce this distortion caused by the nonlinearity of the emitter-base circuit. If a small unbypassed resistor is added in series with the emitter resistor, as shown in Fig. 7-2, the amount of distortion will be reduced.

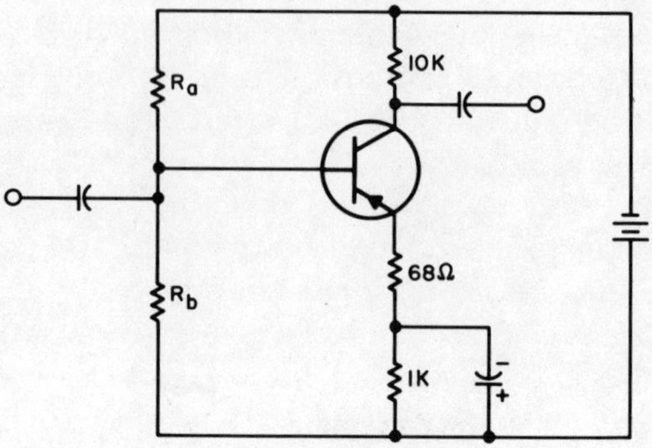

Fig. 7-2. Emitter feedback.

7.1 Effects of Feedback

If $h_{ib} = 30$ ohms, which is the value for $I_E = 1$ mA, the formula for voltage gain now becomes:

$$A_e \approx -\frac{R_L}{h_{ib} + R_E} = -\frac{10K}{30 + 68} \approx -100$$

The 1K resistor is for the improvement of bias stability (notice that it is bypassed for ac).

The voltage gain is reduced by a factor of about 3 as a result of the added resistor, but the voltage waveform distortion is reduced by nearly the same amount. Without feedback,

$$A_e \approx 0.9 \times 10K/30 \approx 300$$

Therefore, the amount of feedback is

$$20 \log 100/300 = -20 \log 3 = -20(0.48) = -9.6 \text{ dB}$$

In the previous example, feedback was applied to a single stage. More typically, feedback is applied to several stages or to a complete amplifier. Using feedback, there can be considerable variation within the individual stages, but, if the open loop gain is sufficient, the gain with feedback can be held constant. In addition, the gain with feedback can usually be made equal to the ratio of two resistors.

Take, for example, the problem of building an amplifier with a voltage gain of 1000, an input impedance of 1 megohm and an output impedance of 50 ohms. The easiest and probably the best way to build this amplifier is to use feedback to set the gain at 1000 and to use an emitter follower to achieve the 1-megohm input impedance.

Referring to Fig. 7-3 and the rule-of-thumb formulas in Appendix 1, we select the values as follows: Emitter follower Q_2 is a buffer stage that is used to prevent the relatively low input impedance of Q_3 from appearing in shunt with the emitter resistor of the input emitter follower. The input resistance at the base of Q_2 is approximately equal to the product of h_{fe} and the input resistance of Q_3. Noting that R_i for stage 3 is about 5K, the value of R_i for Q_2 is

$$R_i \approx 0.8 \times 100 \times 5K$$
$$\approx 400K$$

This 400K resistance will have essentially no effect on the 20K emitter resistor of Q_1.

The value of R_i for Q_1 can now be calculated.

$$R_i \approx 0.8h_{fe} \times 20K$$
$$\approx 1.6 \text{ megohm}$$

The 1.6-megohm R_i in parallel with the 2.2-megohm base bias resistor gives an input impedance of better than 925K and this is close enough to the desired input impedance of 1 megohm.

Stages Q_3 and Q_4 are used as the gain stages. The open loop gain is A_3A_4 and can be quickly calculated by using the rule-of-thumb formulas in Appendix 1. The bias resistors for the base of Q_3 are adjusted to produce 1-mA d-c emitter current. The resistor values are calculated in exactly the same way as demonstrated previously. The emitter resistor in Q_4 is made the same value as the collector resistor in Q_3. Assuming 1-mA of current flows through the Q_3 collector resistor, almost exactly the same value of current passes through the emitter resistor of Q_4. Referring to Fig. 7-3, the open loop gain, A_{OL} (gain without feedback), is

$$A_{OL} \approx \left(\frac{-0.9R_{L3} \parallel R_{i4}}{h_{ib} + R_E}\right)\left(\frac{0.9R_{L4}}{h_{ib}}\right)$$
$$\approx \left(\frac{0.9 \times 10K \parallel 2.4K}{63}\right)\left(\frac{0.9 \times 5K}{30}\right)$$
$$\approx 4300$$

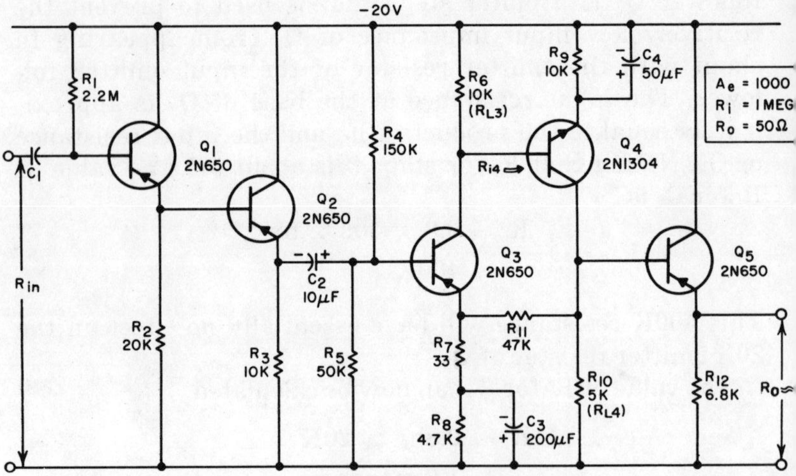

Fig. 7-3. Four-stage amplifier with feedback.

The closed loop gain (gain with feedback) is given by the formula:

$$A_{CL} = \frac{A_{OL}}{1 + \beta A_{OL}} = 1000 = \frac{4300}{1 + 4300\beta} = \frac{1}{\frac{1}{4300} + \beta}$$

or
$$\frac{1000}{4300} + 1000\beta = 1$$

So
$$\beta = \frac{1}{1K} - \frac{10}{43K} = \frac{33}{43K}$$

where β is the feedback factor.

The feedback resistors in Fig. 7-3 are R_7 and R_{11}. A value of 47K for R_{11} gave an overall measured gain of just over 1000. The feedback resistor should be made slightly larger than the value calculated to compensate for the slight losses in the two emitter followers.

The output impedance R_o of Q_4 is considerably lower than the 5K load resistor as this stage uses collector feedback. Calculations yield an R_o of approximately 600 ohms and the measured results give an effective R_o of about 800 ohms. An emitter follower is used to reduce this value to the desired 50 ohms output impedance. The expression for R_o of an emitter follower is

$$R_o \approx \frac{R_g + h_{ie}}{h_{fe}}$$

Note that with an emitter follower, the output impedance is partially dependent upon the generator impedance, R_g. Since $h_{ie} \approx h_{fe}h_{ib}$, the above expression reduces to:

$$R_o \approx \frac{R_g}{h_{fe}} + h_{ib}$$

The value of R_g for Q_5 is about 800 ohms; i.e., the output impedance of Q_4. If $h_{fe} \approx 100$, the only unknown quantity is h_{ib}. (The value of R_o is given as 50 ohms on the illustration.) Now, solving for h_{ib},

$$h_{ib} \approx R_o - \frac{R_g}{h_{fe}}$$

$$\approx 50 - \frac{800}{100}$$

$$\approx 42 \text{ ohms}$$

The base voltage of Q_5 is approximately 5 volts (the voltage on the collector of Q_4) and the emitter current should be about 0.75 mA to produce the desired h_{ib} of 42 ohms. Remember, h_{ib} varies inversely with current and has a value of 30 ohms at 1 mA of emitter current. Thus,

$$R_E \approx 5/0.75$$

$$\approx 6.8\text{K}$$

Using a 6.8K resistor for R_E, we obtain a measured R_o of 47 ohms. This is close enough to the desired value.

The frequency response of the amplifier in Fig. 7-3 is from 17 Hz to 200 kHz.

CHAPTER 8

Other Semiconductor Diodes

8.1 Breakdown Diodes

After the reverse-breakdown phenomenon of diodes was discovered, Dr. Zener proposed a theory for their operation. Zener proposed that as the electric field (potential) increased, a point was reached where this field was strong enough to pull electrons from the lattice of an atom and place them in the conduction band. This action would cause a sudden increase in current and a large reduction in diode impedance. Zener also reasoned that this type of breakdown should exhibit a negative T.C. (temperature coefficient).

As time went on and a better understanding of semiconductor materials became available, the Zener theory, as applied to the breakdown phenomenon, was replaced by the avalanche breakdown theory. Avalanche breakdown is analogous to the Townsend gas discharge. As the electric field increases, the velocity of those electrons in the conduction band increases to the point where they have enough energy to knock other electrons out of the lattice structure. The electrons that are knocked out of the lattice may gain enough energy before they recombine to repeat the process. When these conditions are met, the current *avalanches*.

Interestingly enough, it can be shown that avalanche breakdown should exhibit a positive T.C.

Very low voltage breakdown diodes probably break down from true Zener discharge. In line with the Zener theory, these diodes exhibit a negative temperature coefficient. Diodes that break down at higher values of voltage probably do so primarily by the avalanche process, and these diodes exhibit a positive temperature coefficient. There are also some diodes that exhibit essentially a zero temperature coefficient. These diodes break down in a transition region of 4.5 to 6 volts and, in these cases, the breakdown process is partly avalanche and partly by Zener action.

Figure 8-1 shows a circuit that is used quite often in power supplies. The 11.5-V diode is an avalanche breakdown diode and thus has a positive (+) T.C. The forward-biased (temperature compensation) diode, as was pointed out in an earlier chapter, has a negative (—) T.C. This combination provides good stability by compensating for changes in temperature.

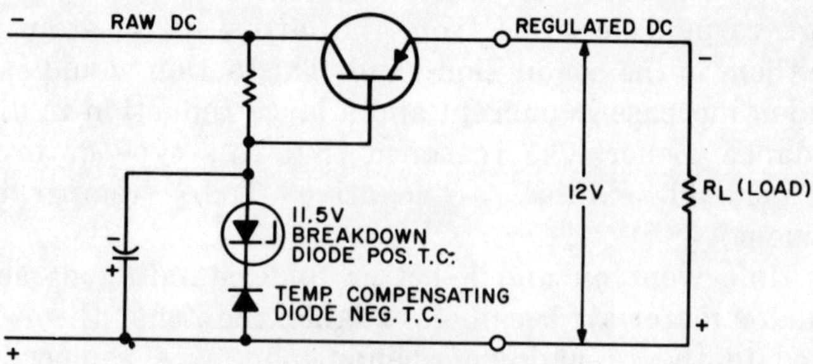

Fig. 8-1. Avalanche-diode temperature compensation.

8.2 Step Recovery Diode

This diode is similar to the PN junction diode, but achieves its unique properties of long charge storage and fast switching through special doping techniques in the vicinity of the junction. The important concept here is the treatment of stored minority carriers.

The action in a step recovery diode begins when the driving waveform, Fig. 8-2A, reverse biases the diode. Just prior to reverse bias, the forward current caused substantial amounts of minority carriers to be placed on either side of the junction. Through heavy doping of the region near the junction, the minority carrier storage does not extend very far out into the semiconductor materials. Because of this restriction of stored charge, when recombination ends, it ends abruptly.

A good analogy here is a water faucet and a barrier. See

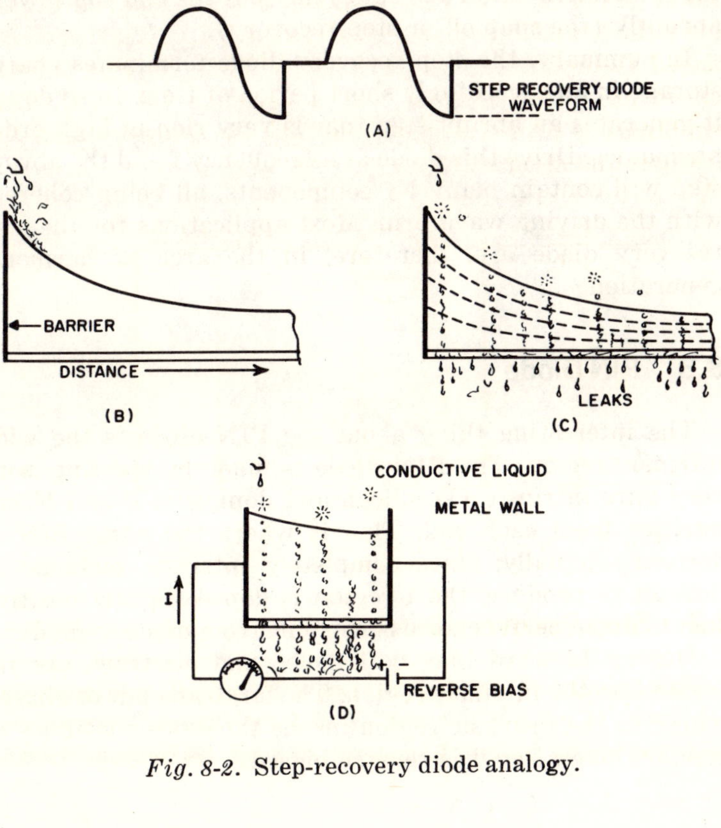

Fig. 8-2. Step-recovery diode analogy.

Fig. 8-2B, which describes a normal PN junction diode. As long as the faucet is on (forward current), the liquid level falls exponentially with distance from the barrier. When the current is turned off, recombination begins to remove the stored carriers. This idea can be simulated by putting holes in the bottom of the container, as shown in Fig. 8-2C.

The dashed lines in Fig. 8-2C show that the liquid drains out gradually. The only difference between this analogy and the action of a step recovery diode is that the high doping level near the junction must be simulated by putting a restricting wall in the picture. This is done in Fig. 8-2D.

Now when the faucet is turned off, i.e., reverse biased, the conductive fluid runs out evenly but the end comes very abruptly (the snap off or step recovery).

In summary, the step recovery diode terminates charge storage in an exceedingly short period of time. In so doing, it generates an abrupt step that is very rich in high order harmonics. Drive this diode at a frequency f and the abrupt step will contain many Nf components, all being coherent with the driving waveform. Most applications for the step recovery diode are, therefore, in the area of harmonic generation.

8.3 PIN Diode

The interesting thing about the PIN diode is the wide intrinsic region. The PIN diode is made by starting with very pure intrinsic (I) silicon and doping in P and N impurities from each end. This is where the name PIN is derived. Actually, since completely intrinsic material is difficult to produce, the I region is doped slightly positive just to make performance consistent from diode to diode.

During forward bias both holes and electrons are injected into the intrinsic region. The large amount of charge stored in the intrinsic region means the diode continues to conduct long after it is reverse biased.

The water analogy that was useful in explaining the step recovery diode is also useful here. Of course, recombination goes on at a fairly rapid rate as both holes and electrons are injected into the intrinsic region. Recombination is illustrated in Fig. 8-3 by the holes in the bottom of the container. When the driving signal reverses, the large

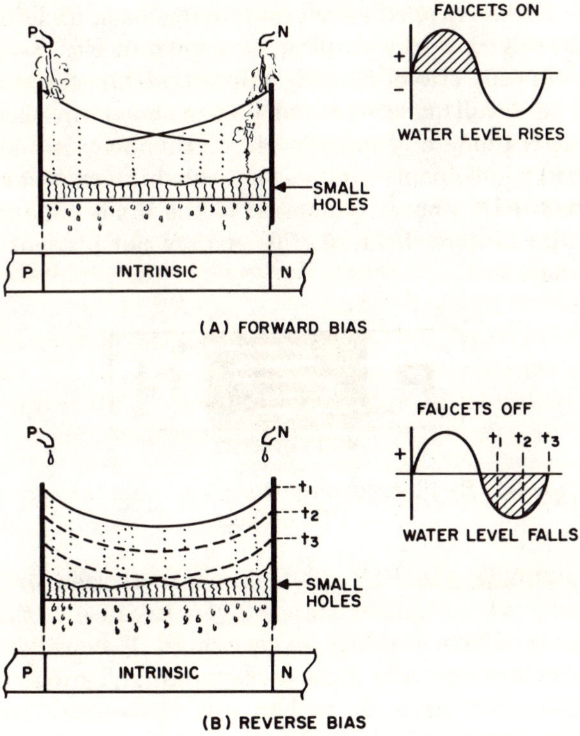

Fig. 8-3. PIN-diode analogy.

amount of stored charge causes the diode to continue to conduct for a relatively long time. This is the key to its application. A microwave signal, for example, keeps the diode *on* throughout the entire reverse-bias cycle. As a result, the PIN diode is never *off* and it does not really behave like a diode, but rather as a variable resistor. The amount of resistance is a function of the d-c forward bias, i.e., more forward bias produces more stored charge and, therefore, lower d-c resistance.

From a microwave point of view, the wide intrinsic region of the PIN diode means that during the back bias period of the r-f cycle, carriers are trapped in the intrinsic region. These trapped carriers, moving back and forth for several cycles before escaping, as shown in Fig. 8-4, represent a current traveling through a resistance for a finite time. The resulting dissipated power shows up as heating of the PIN diode and getting rid of this heat is one of the real design problems in making the higher, microwave-frequency PIN-line attenuators. If the diode mount is designed to dissipate heat readily, it may not present a good impedance.

Fig. 8-4. Radio-frequency energy passing through a PIN diode.

In summary, the PIN diode is characterized by an exceptionally wide high resistance layer. Because of this wide region, the diode exhibits a very high (several hundred volts) breakdown voltage. Correspondingly, diode capacitance per unit area is rather low. However, forward-bias conductivity can be high because the conductivity of the intrinsic region is increased by the presence of stored charge (conductivity modulation).

The big application for the PIN diode is at microwave frequencies. By providing a variable attenuator that presents a relatively constant match to the microwave source, complex waveform modulation is possible without frequency pulling (i.e., little or no SWR change).

8.4 Hot-Carrier Diodes and the Fermi Level

Although an understanding of Fermi level is not absolutely necessary to study this material, it is being discussed here because it is becoming an increasingly useful concept in explaining some of today's exotic semiconductor devices. Simply put, the Fermi level is the energy level at which there is a 50 percent probability of electron occupancy. This definition fails, however, to convey a graphic description and does not immediately show the significance of Fermi level when two different semiconductor materials are brought together. Let us see, therefore, if we can develop a more meaningful visualization of this important term. Since the Fermi level is an energy level, let us consider it as such and call it a *quasi-mean* energy level, which gives an indication of the energy condition of the electrons in a material. The important thing to note here is that when two different Fermi level materials are brought together (with no external bias), there can be only one resultant Fermi level. This is just like two different water levels suddenly being brought into contact with each other.

There is a fairly good analogy between liquid levels and Fermi levels as we can see from the following example. See Fig. 8-5. Consider the partition between the two liquids as being made in two parts—one wall with a hole in the bottom and the other a plain solid barrier. If the solid barrier is removed, Fig. 8-5B, the two liquids flow through the hole and settle at a uniform level. (It is intuitively obvious, but the reason is because a system in equilibrium seeks the lowest possible energy level.) At the

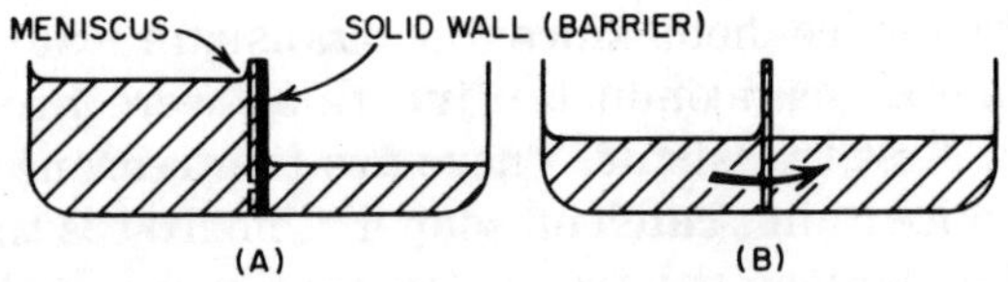

Fig. 8-5. Fermi level water analogy.

surface of the water and along the partition a meniscus will form, with the height of this meniscus proportional to the liquid surface tension. This analogy is a useful one as the meniscus is comparable to the barrier that is created in *hot carrier diode*. A hot carrier diode has a metal N junction and our water analogy is useful in explaining its operation.

Consider a hot carrier diode with the Fermi level in the semiconductor above that in the metal. This is pictured in Fig. 8-6A. The metal can be compared to a low surface tension liquid, whose meniscus is virtually nonexistent (this can be explained by the high number of electrons available in the metal, i.e., the valance and conduction bands overlap at room temperature). For an electron to be moved

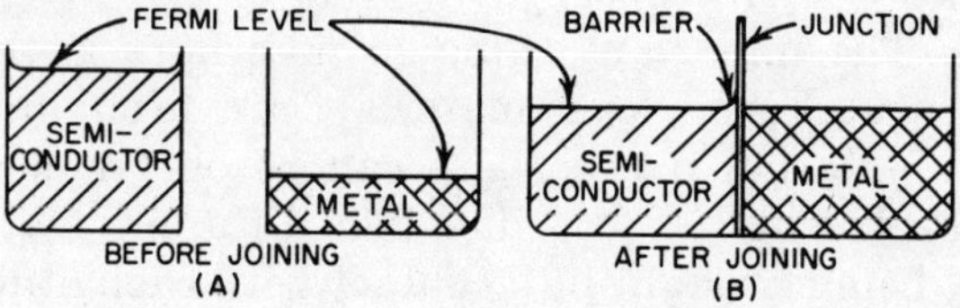

Fig. 8-6. Formation of hot-carrier diode barrier.

from the N-doped semiconductor into the metal, enough forward bias must be applied to get the electron up over the barrier (meniscus). When the electron clears the barrier and flops over into the metal, Fig. 8-7, it will momentarily have more kinetic energy (i.e., be hotter) than those electrons in the metal. This condition only lasts for about 100 femtoseconds, i.e., 0.1 picoseconds.

The hot-carrier diode is the most exciting entry in the solid-state device field since the transistor. Surprisingly enough, metal semiconductor junctions were around long before the first transistor. There are two reasons why this type of device didn't catch on sooner. The first is that it was the development of the transistor that gave us the manufacturing know-how necessary to build a reliable metal

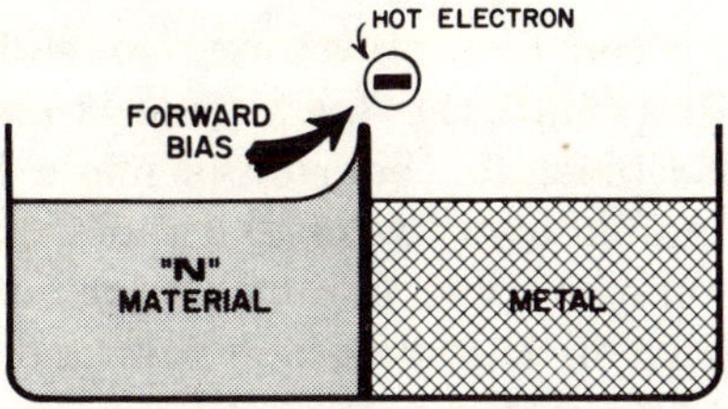

Fig. 8-7. Junction in hot-carrier diode.

semiconductor junction device. The second is that solid-state theory has advanced about tenfold since Shockley and friends put together their first transistor.

What makes the hot carrier diode (and hopefully in the future, the hot carrier transistor) so exciting is the lack of stored minority carriers. It is strictly a majority carrier device—electrons all the way. As you will recall, stored minority carriers must be removed before a diode or transistor can turn off.

The presence of stored minority carriers in the transistor has given us a practical upper-frequency limit of something like 3 GHz. Even to get 3 GHz is exceedingly difficult and quite costly. The hot carrier device with virtually no minority carrier conduction has a theoretical upper frequency limit of at least 60 GHz.

The key to rectification in the metal-to-N junction is that the work function (the amount of energy necessary to remove an electron completely from the control of its parent atom) in the semiconductor is less than in the metal. Thus, it takes less energy to move an electron from the control of the semiconductor atom than from the metal, i.e., the electrons are hotter (have more energy) in the semiconductor than in the metal. When the metal and N-type semiconductor are joined, the energy levels equalize but in doing so, a barrier is created at the interface. During forward bias, the electrons in the N material are pushed toward the barrier. If the forward bias is sufficient (0.3–0.6 volts, depending upon the choice of the metal), the N material electrons are given enough energy to slide up over the barrier and drop into the sea of electrons in the metal. During reverse bias, the barrier is made higher and the electrons are pulled away from the junction.

A comparison of barrier potentials for hot-carrier diodes (HCD) is shown in Fig. 8-8. The forward characteristics are controlled by using different metals.

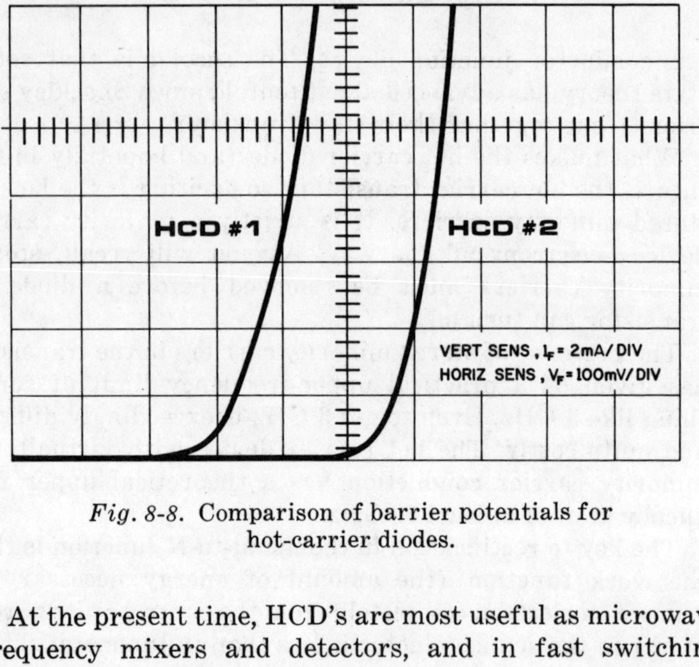

Fig. 8-8. Comparison of barrier potentials for hot-carrier diodes.

At the present time, HCD's are most useful as microwave frequency mixers and detectors, and in fast switching circuits.

8.5 Tunnel Diode

Under conditions of zero bias, the Fermi level in a standard PN junction lies between the conduction and valence bands. This condition is illustrated in Fig. 8-9A. In the tunnel diode, however, exceptionally heavy doping causes the Fermi level to lie inside the respective energy bands, i.e., on the N side the Fermi level lies inside the conduction band and, on the P side, inside the valence band. See Fig. 8-9B. This condition places electrons and holes in close proximity. Typically, the depletion region is only 100 Angstrom units. (1 cm = 10^8 Angstrom units.) This proximity of holes and electrons causes electrons to *tunnel* after nearby holes, and vice versa. Hence the name tunnel diode.

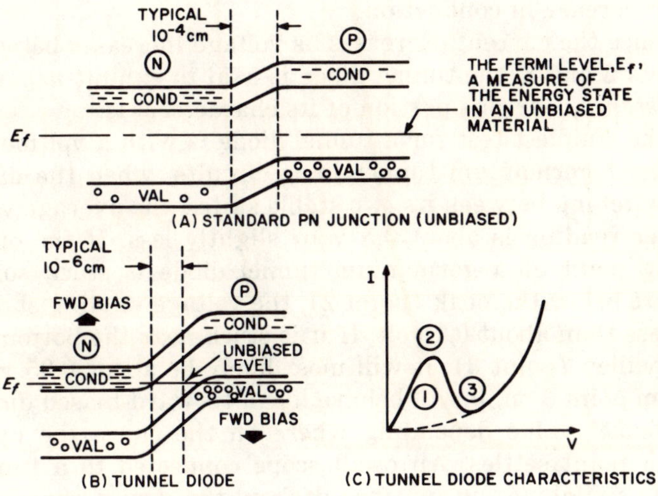

Fig. 8-9. The tunnel diode.

When forward bias is applied to the tunnel diode, two things happen. First, conduction begins immediately because the conduction and valance bands overlap and the charge carriers are pushed toward each other. Second, as the electric field increases, the barrier height (potential hill) is reduced. This causes the conduction band of the N material to move away from the adjacent valence band of the P-type material. See Fig. 8-9B. The peak of the extra current region occurs when the bottom of the N material conduction band is tangent to the top of the P material valence band. This is identified as point 2 in Fig. 8-9C. At this point, a large number of charge carriers have been pushed up to the barrier and the bands are still in a position to allow tunneling. From this point on, conduction decreases rapidly because the two bands begin to move further apart. Thus, tunneling (extra current) stops. At point 3 in Fig. 8-9C, conduction results from normal diode action. The transition between points 2 and 3 occurs rapidly. A further increase in forward bias causes a normal increase in conduction.

Since the current decreases as voltage increases between points 2 and 3, the tunnel diode is said to exhibit *negative resistance* over this portion of its characteristic.

The simplest test for a tunnel diode is with a voltmeter. In most germanium tunnel-diode circuits, when the diode is switching between its two stable states, the average voltmeter reading is about 0.5 V or slightly less. If the operating point of a germanium tunnel diode is stuck somewhere below the peak (point 2), the voltage reading should be less than about 0.1 volt. If it is stuck near the bottom of the valley (point 3), it will most likely be about 0.35 volt. From point 3 on, it will behave like a forward-biased diode, the exact value depending where on the curve the operating point settles. An oscilloscope connected to a tunnel diode switching circuit should show the device switching at its operating frequency. The waveform is very often, but not necessarily, a square wave.

For testing tunnel diodes out of circuit, the test setup of Fig. 8-10 can be used. Actually this test setup will display the V-I characteristics of just about any diode. The current is monitored across the 1K resistor so the displayed

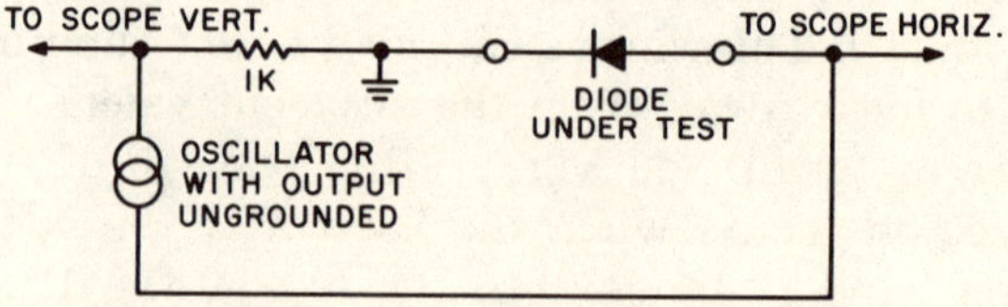

Fig. 8-10. Diode test circuit.

voltage equals the diode current in milliamperes. For oscilloscopes with a common ground, the vertical channel should be able to display a *negative up* signal. Otherwise, the display is inverted.

CHAPTER 9

Special Devices

9.1 The Silicon Controlled Rectifier

The silicon controlled rectifier (SCR) is the solid-state equivalent of a thyratron. A simple SCR circuit is shown in Fig. 9-1. The device conducts when a positive pulse is applied to the gate electrode.

Perhaps the easiest way to analyze an unknown device is to assume an *on* or *off* state and then prove the assumption true or false. In the case of the SCR shown in Fig. 9-1,

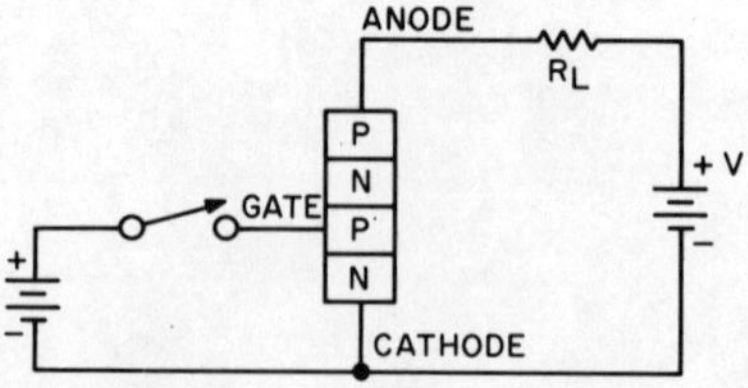

Fig. 9-1. The silicon controlled rectifier (SCR).

let us assume the device is conducting. If it is, there must be a voltage drop across the device. This voltage becomes progressively more positive moving from cathode to anode. The resultant voltage gradient makes the top and bottom PN junctions forward biased and the middle PN junction reverse biased. Since these conditions cannot exist, the first assumption was wrong and the device is *off*.

To understand the mechanism of operation, it is helpful to consider the SCR as two separate transistors. This concept is illustrated in Fig. 9-2.

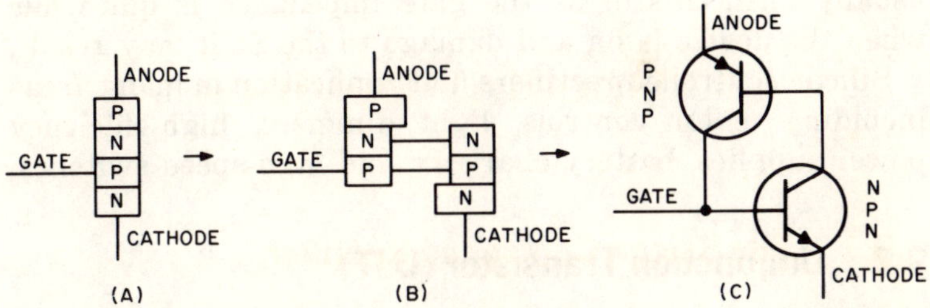

Fig. 9-2. Equivalent circuit of SCR.

In operation, the SCR resembles a positive feedback amplifier. Once conduction begins it continues and makes both transistors saturate. Why, then, doesn't the device get started by some small noise or leakage current? The answer lies in the behavior of h_{FE} with current. See Fig. 9-3. Normal leakage current is so low that the combined h_{FE} of both equivalent transistors gives less than unity gain in the two-transistor feedback amplifier. However, when a momentary, positive pulse is applied to the gate circuit, the effective h_{FE} is momentarily greater than unity and the

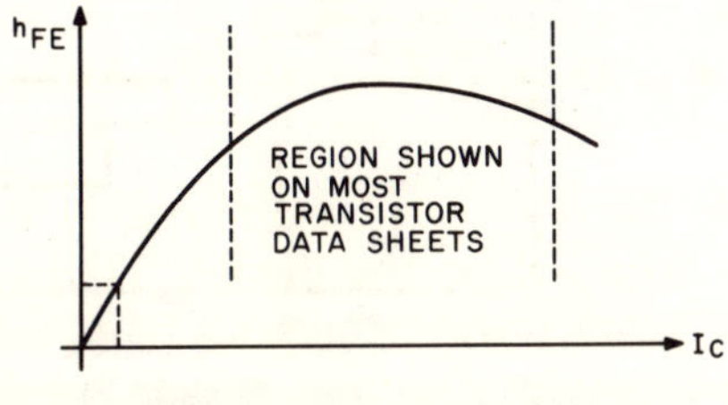

Fig. 9-3. Dependence of h_{FE} on I_C in a silicon transistor.

two equivalent transistors saturate. Once saturated, current through the device is enough to keep the combined h_{FE} greater than unity.

About the only way to turn off an SCR is to reduce the gate-to-anode current to less than the *holding current* (i.e., where the combined h_{FE} is less than unity). Attempting to put a negative pulse on the gate for *turn off* purposes is usually unsuccessful as the gate impedance is quite low when the device is on and damage to the SCR may result.

Silicon controlled rectifiers find application in many areas including motor controls, light dimmers, high-efficiency power supplies, battery chargers, and high-speed switches.

9.2 Unijunction Transistor (UJT)

The UJT is best explained with an example. Consider the circuit of Fig. 9-4. When switch S is closed, capacitor C_1 starts charging from zero toward +V. (This assumes no prior charge on C_1.) During this charging time, there is a uniform voltage drop along the bar of N material. Being a

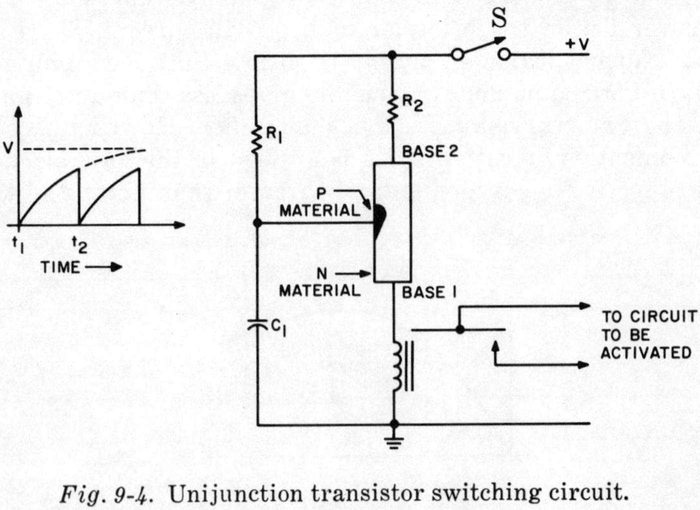

Fig. 9-4. Unijunction transistor switching circuit.

semiconductor, the N-material bar behaves like a medium value resistor (5K to 10K) with a voltage gradient along its length.

From time t_1 until time t_2, the voltage on the P-material emitter (voltage across C_1) is less positive than the voltage in the surrounding N material. This produces a reverse-biased PN diode. As the voltage on the capacitor continues to rise toward +V, a critical value is reached where this PN diode becomes forward biased and, at that time, electrons are injected into the P-material emitter. More significant, however, holes from the P emitter are injected into the N-material bar.

The presence of a large number of holes in the silicon bar greatly lowers its impedance. As a matter of fact, the emitter-to-Base 1 region behaves like a forward-biased diode. The energy on C_1 now discharges into the Base 1 load circuit. The change of state occurs at time t_2 in Fig. 9-4.

After capacitor C_1 has lost its charge, the field from Base 2 takes over control, reestablishing the voltage gradient along the N-material silicon bar and reverse biasing the emitter-Base 1 PN junction. The cycle described above now repeats unless there is some turn-off mechanism in the load. A set of relay contacts is often used for this purpose when the unijunction transistor is a relay used as a time delay.

Applications of the UJT include sawtooth generators, time-delay circuits, staircase waveform generators, and pulse-forming circuits for firing SCR's.

When replacing a UJT, care should be taken not to reverse the connections to the two bases. Although no damage should result, since the bases are simple ohmic connections, the device is not symmetrical. The P-material emitter is closer to Base 2 than to Base 1. The result of misconnecting Base 1 and Base 2 would be to make the UJT inoperative.

9.3 The Field-Effect Transistor (FET)

The field-effect transistor (FET) combines some of the best properties of the vacuum tube with those of the transistor. Like the vacuum tube, it is a voltage-operated device with very high input impedance. Like all transistors, it has no filaments or microphonics, can operate at relatively low voltages, and can have either holes or electrons as the current carriers.

There are two popular types of field-effect transistors in use today: the junction-gate and the insulated-gate. The junction-gate will be considered first as its use is now more widespread. However, the advent of integrated circuits may spur the use of the insulated-gate device, since it can be used without load resistors in some logic circuits. The d-c requirements may also be simplified.

In its simplest form, the operation of a junction-gate field-effect transistor can be compared to a hand squeezing a garden hose. A typical field-effect transistor is constructed by taking a bar of silicon (the hose), doping it (either N or P), and adding ohmic contacts at each end. Next, a section or layer of P (or N) material (the hand) is built into the middle of the bar. When reverse bias is applied between this section of P (or N) material (called the gate) and the bar (called the channel), control of the carriers in the channel can be obtained. A simple circuit arrangement for an N-channel FET is shown in Fig. 9-5.

Compared to a conventional pentode vacuum tube, the drain is equivalent to the anode (plate), the source is equivalent to the cathode, and the gate is equivalent to the grid. Like the conventional vacuum tube, the junction-gate FET is basically an *on* device and must be biased part-way off to set the operating point in a position for amplification. See Fig. 9-6.

Referring to Fig. 9-7, the normal operating region for amplification is past the knee (pinch-off) voltage. Pinch-off voltage is the reverse bias applied between the gate and drain and causes the depletion region of this PN junction to restrict the flow of carriers through the source-drain channel.

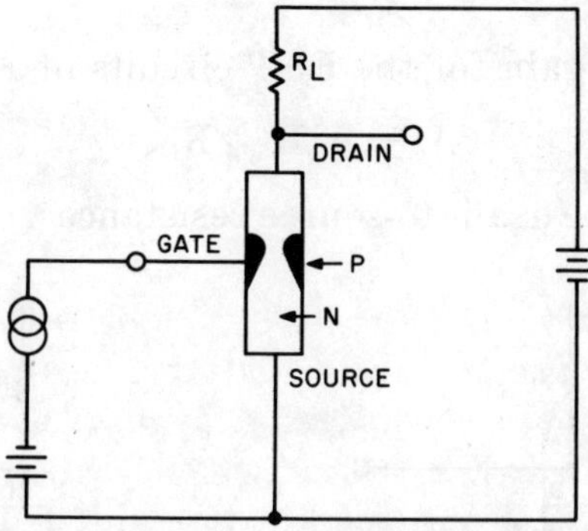

Fig. 9-5. An N-channel FET.

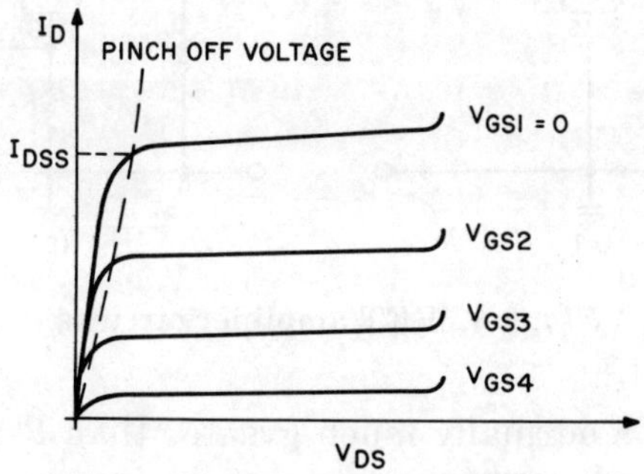

Fig. 9-6. FET characteristics.

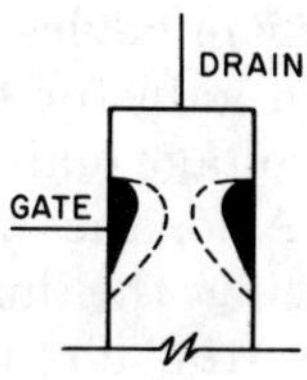

Fig. 9-7. Depletion region around an FET gate.

Reverse bias on the gate-channel junction is greatest near the drain and is the reason for the tapered shape of the depletion region (shown dashed) around the gate in Fig. 9-7.

The voltage gain for the FET circuits of Fig. 9-8 is

$$A_e = g_m R_L \parallel R_{DS}$$

where R_{DS} is the drain-to-source resistance.

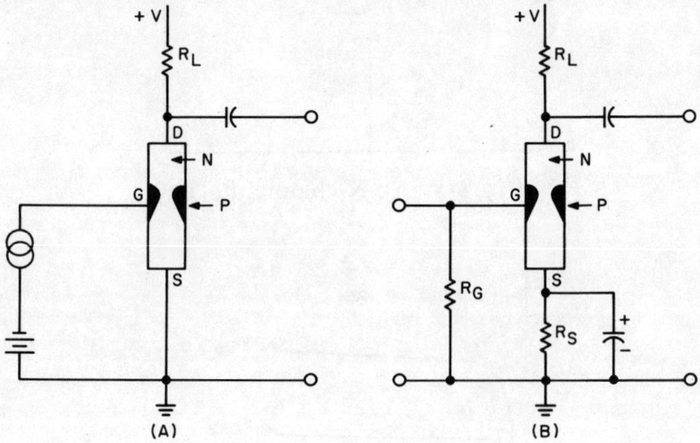

Fig. 9-8. FET amplifier circuits.

Since R_{DS} is normally much greater than R_L, the expression usually reduces to:

$$A_e \approx g_m R_L$$

The term g_m, which can be found on the data sheet, varies both with current and from device to device. The result is considerable variation in gain for the circuits of Fig. 9-8.

The expression for voltage gain is the same as for a pentode vacuum tube. Also, the input impedance of the device is very high, being the impedance of a reverse-biased diode; therefore, the circuit input impedance in Fig. 9-8B is

$$R_i \approx R_G$$

The acceptance of the FET has been quite slow because the g_m of the device is relatively low and voltage gains are, for the most part, below those obtainable with the more conventional type of transistor. In addition, FET's tend to have greater amounts of high-frequency noise but less low-frequency noise. However, FET's do have one big advantage over conventional transistors; viz., they are much less susceptible to radiation damage because they are strictly majority-carrier devices; that is, the carriers stay in the same kind of material when passing from source to drain.

For troubleshooting, there are two simple tests that can be made when FET's are out of the circuit. See Fig. 9-9. In the circuit of Fig. 9-9A, the drain and source are connected to each other. Reverse bias just below the breakdown value is applied between the gate and the source-drain connection. The resultant leakage current should be below I_{GSS}, the maximum drain-to-gate leakage current. Typical values for I_{GSS} are usually between 0.01 and 1.0 mA at room temperature.

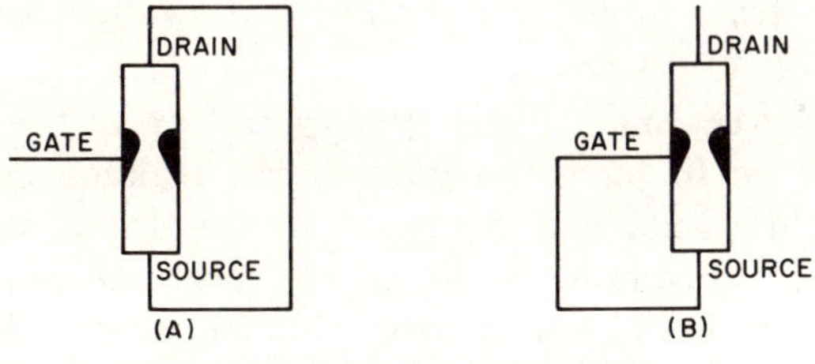

Fig. 9-9. An FET test circuit.

In the circuit of Fig. 9-9B, bias is applied between the drain and the gate-source common connection. The resultant current should be between I_{DSS} max and I_{DSS} min, but definitely above I_{DSS} min. I_{DSS} is the drain-to-source current with zero volts between the gate and the source.

This is the current the FET can pass without forward biasing the gate diode. Shorting the gate to the source in an operating junction-gate, field-effect transistor will remove the pinch-off effect and cause the transistor to go into heavy conduction.

In addition to the above tests, the drain-to-source resistance can be measured. Typical values for junction-gate devices are between 500 ohms and 5K. When making this test, it is best to connect the gate to the source. Leaving the gate disconnected allows a charge to build up on the gate, causing the measured resistance to fluctuate. Since the FET is similar to the vacuum tube, an open gate has a very high impedance and behaves just like an open grid. Care should be taken to connect the ohmmeter properly. In a P-channel unit, the negative lead should be connected to the drain. In an N-channel unit, the positive lead should be connected to the drain. If the ohmmeter leads are reversed, the readings will not be the same when the ohmmeter scale is changed. This is because the gate-drain becomes a forward-biased diode when the leads are connected improperly.

Other troubleshooting tests include measuring the forward and reverse diode characteristics of the gate-source and gate-drain junctions. Values are similar to the good quality silicon diodes.

Probably the most likely type of failure in the junction-gate FET is to have the gate-source leakage current increase and produce a decrease in the input impedance. High-input impedance is one of the most useful characteristics of the FET, and having this impedance lowered is undesirable. The change in input impedance is often hard to detect. Simple ohmmeter checks usually do not detect this leakage change, which may be in the nanoampere region. However, the leakage may be external to the FET; therefore, before replacing a leaky FET, try cleaning the gate-to-source, gate-to-drain, and gate-to-ground connections. This type of external leakage is more prominent on printed circuit boards than with point-to-point wiring because the flux normally used in printed-circuit work is both corrosive and slightly conductive. With proper cleaning, the residual flux usually can be removed.

Much of what has been said about the junction-gate device applies equally well to the insulated-gate, field-effect transistor. This device has a voltage gain, which is essentially equal to the $g_m R_L$ product, and a high input impedance. Moreover, it has a source, gate, drain, and channel. At this point, however, the similarity ends. As the name implies, there is a thin film of insulation between the gate and channel of the insulated-gate FET. See Fig. 9-10. The gate and channel electrodes act like the plates of a capacitor and the insulation acts like the dielectric. As a result, a positive charge on the gate of an N-channel device drives electrons into the channel thereby increasing conduction.

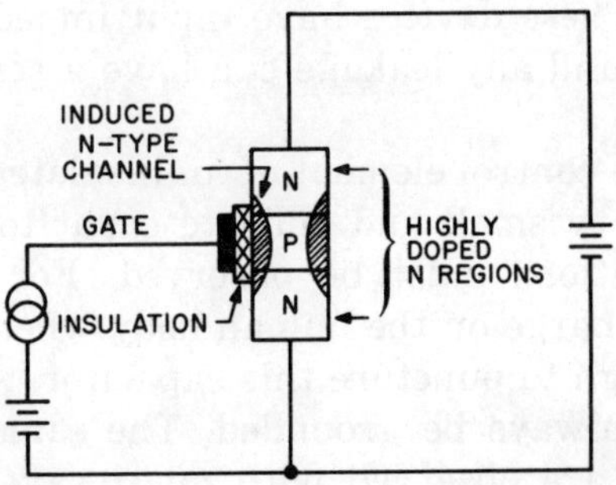

Fig. 9-10. The MOSFET biased for enhancement mode.

The channel may be made in either of two forms; viz., high or low conductivity. With a high-conductivity channel, operation is similar to the junction-gate device. To behave as an amplifier, therefore, the bias point is adjusted for reduced conduction. With a low-conductivity channel, the bias is adjusted to enhance conduction, and this type of operation is often called *enhancement-mode operation.*

Insulated-gate units should exhibit even better resistance

to radiation damage than do junction-gate, field-effect devices. However, radiation often causes ionization along the gate silicon dioxide layer which causes performance deterioration. The insulated-gate FET also finds application in the digital field where its exceptionally low leakage currents and ability to provide direct-coupled inversion without level shifting between stages are advantageous.

Because enhancement-mode, insulated-gate FET's are field operated devices and have insulation under the gate, they do not lend readily to forward/reverse bias diode checks. However, some of the troubleshooting techniques that are effective on conventional PNP and NPN transistors can be applied. A good in-circuit check, for example, is to short the gate and source to remove the enhancing field and turn off the device.

The previous comments about leakage on printed-circuit boards are especially true when working with insulated-gate FET's. These devices have input impedances as high as 10^{15} ohms and any leakage can have a serious effect on performance.

Because the control element of an insulated-gate FET is, in effect, a very small and delicate capacitor, special servicing precautions must be observed. For example, the electrostatic charge on the human body when the humidity is low is enough to puncture this capacitor. Also, soldering irons should always be grounded. The same general precautions that are observed with microwave point-contact mixer and detector diodes should be observed when working with insulated-gate, field-effect transistors.

9.4 Testing FET's

By way of review, the three types of FET's are the Junction Gate, the Insulated Gate (non-enhanced or depleted), and the Insulated Gate (enhanced). The gate insulation (see Fig. 9-10) is usually a metal oxide, thus the name metal-oxide semiconductor, or MOS. The similarities and

TABLE 9-1. PRINCIPAL CATEGORIES

The Three Basic FET's	Junction Gate (J-FET)	Insulated Gate—Non-enhanced (DEPLETION) (MOSFET)	Insulated Gate—Enhanced (MOSFET)
Type of Channel	Can be either N or P channel	Can be either N or P channel	Can be either N or P channel
Condition of channel	High conduction (ON at zero bias)	High conduction (ON at zero bias)	Low conduction (OFF at zero bias)
Comments:	Similar to vacuum tube *pentode*	Insulation is between gate and channel, comparable to a delicate capacitor. Input resistance can be as high as 10^{15} ohms. Also similar to pentode.	

differences among the three basic types of FET's are described in Table 9-1.

The FET symbols and a schematic equivalent of what is actually inside the devices are shown in Fig. 9-11. Also shown is why FET's are hard to test. For example, a dc ohmmeter could be used to check the J-FET diode but

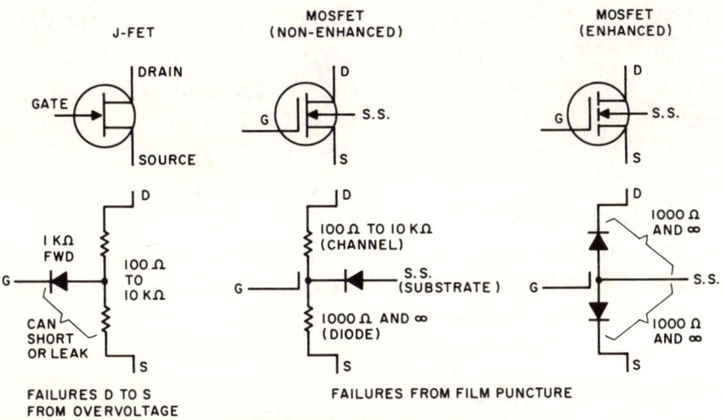

Fig. 9-11. FET symbols and schematic representation (examples are all N channel).

would not work on a MOSFET device due to gate insulation.

In addition, a non-enhanced MOSFET will show a resistance from source to drain while an enhanced MOSFET will not. However, between substrate and source or drain in both MOSFET's you will essentially see a diode. Incidentally, *enhanced* as used here means a channel must be "created" by the gate field to allow carriers (electrons in N channel, holes in P channel) to move from the source to the drain. Therefore, in an enhanced MOSFET you would have to *forward bias* the gate/source to start conduction. Thus you can see that enhanced MOSFET's are basically *OFF* devices, while non-enhanced (depleted) MOSFET's and J-FET's are basically *ON* devices.

Testing FET's—In Circuit

Before removing any FET from its circuit, try these in-circuit test tips: *J-FET*—short the gate to source and the device should start conducting heavily, as you have removed the field which pinches the channel. You can check this by monitoring either the dc source voltage (it will go up) or the dc drain voltage (it will go down). *MOSFET Non-enhanced*—short the gate to source, or gate to substrate. Usually you will find the source and substrate connected together. As with the J-FET, the non-enhanced MOSFET will conduct more heavily. *MOSFET Enhanced*—short the gate to source, or gate to substrate. As mentioned above, the source and substrate are usually connected together. In this case the device should turn off, since the enhancing field has been removed. Check by monitoring the dc voltages on the source or drain.

Testing FET's—Out-of-Circuit

For an out-of-circuit tester to be really useful, it must be able to test both P and N junction-gate and insulated-

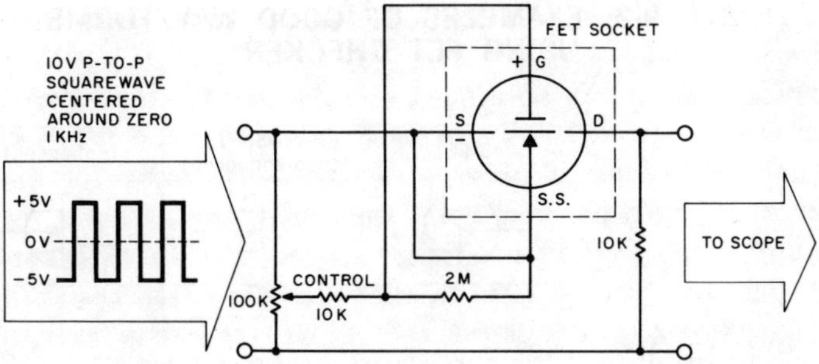

Fig. 9-12. FET checker.

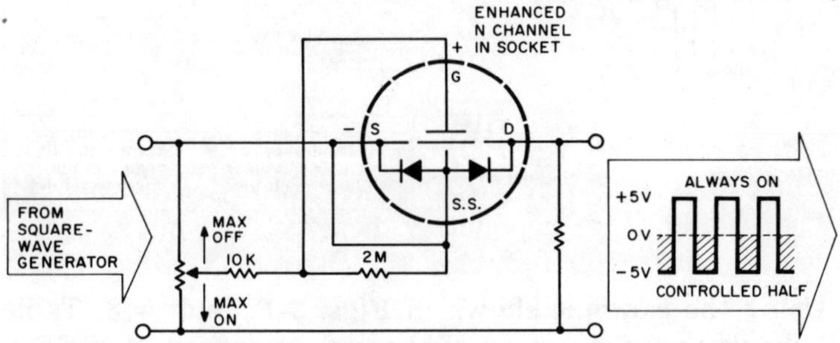

Fig. 9-13. Tester redrawn to show FET diode equivalent and how control affects output.

gate FET's, plus have the ability to test enhanced as well as non-enhanced FET's. Figure 9-12 shows the schematic of a FET checker that will test all of these types, including the enhanced MOSFET in the test socket.

This is a dynamic checker that uses a 1-kHz square wave to provide both signal and "power" to the device under test. Table 9-2 describes the waveforms you should see on an oscilloscope when testing good FET's.

TABLE 9-2. EXAMPLES OF GOOD WAVEFORMS USING FET CHECKER

	J-FET	MOSFET (non-enhanced)	(enhanced)
	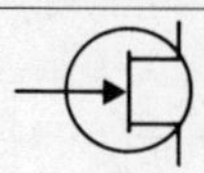	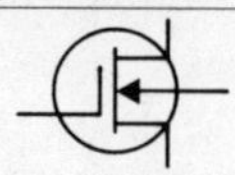	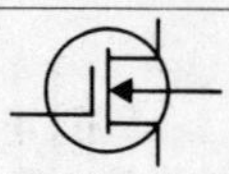
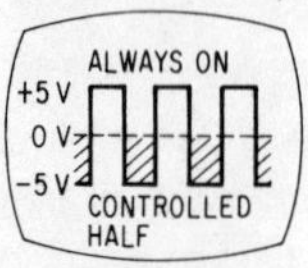	P channel	P channel	N channel
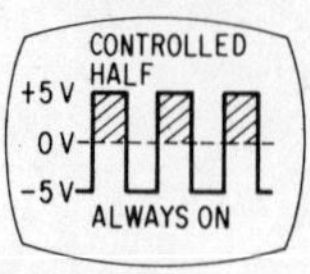	N channel	N channel	P channel
	3 Leads (4 with case) controlled half does not go all the way to zero	4 Leads controlled half does not go all the way to zero	4 Leads controlled half goes all the way to zero

Using the example shown in Figs. 9-12 and 9-13, Table 9-2 shows that for a good N-channel, enhanced MOSFET, the control should not affect the top half of the squarewave but should make the bottom half vary in value from normal, that is from —5V, to zero. (Now you can see that your squarewave input signal must swing between a positive voltage and a negative voltage to cover all FET types.) To make an N channel enhanced MOSFET conduct, we have to put a positive signal on the gate with respect to the source. This happens when the bottom of the potentiometer is more positive than the top of the pot, that is, whenever the squarewave input is in the region below 0 volts. This is the region where you would have control because by varying the position of the pot wiper you vary the amount of *enhancement* voltage applied to the source.

The reason the top half of the squarewave is always on is because the FET diode between substrate and drain is forward biased during this time. The diode between substrate and source is shorted out by the jumper wire on the tester socket. This jumper provides a reference point for the substrate so it won't float around. While we're on that subject, the 2-meg resistor provides a bleed path for any charge that might build up on the gate if the wiper accidentally lifts off the pot. The 10K resistor is a current limiter needed only for J-FET's and then only when the pot wiper is at the maximum to forward bias the gate/source diode. Without the 10K limit resistor you would see amplitude loading on the squarewave generator. The important thing about the generator is that the output be symmetrical around zero and that you can get a 10V peak-to-peak output.

You can verify Table 9-2 by doing a similar analysis for a J-FET and a non-enhanced MOSFET. The oscilloscope display should look like what is shown in Table 9-2, plus if you have leakage in the FET you may have rounding on the trailing edge of the squarewave. Obviously, with

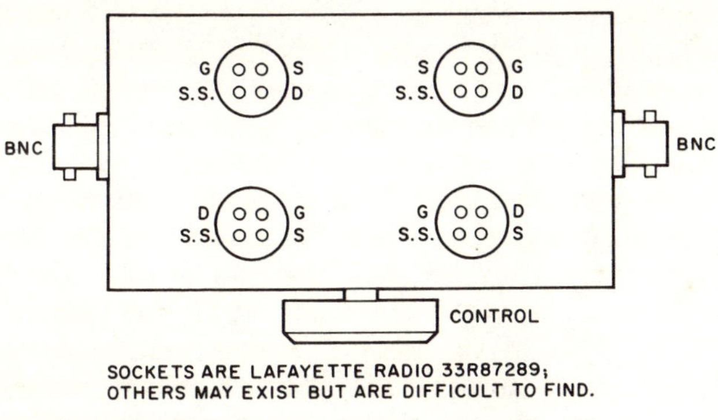

Fig. 9-14. Top view of FET tester.

a completely defective or shorted gate, you won't get any gate action so you won't get any change when you vary the control. After you have built your checker, you had best experiment to learn the subtleties of what represents good and bad.

When you begin constructing a checker of your own, I would recommend installing four 4-pin sockets in parallel to cover the many different FET socket configurations. The ones I recommended are shown in Fig. 9-14. Once you have these configurations identified, you can use the checker, along with Table 9-2, to identify unknown FET's.

CHAPTER 10

Handling Transistors

Several of the early taboos in working with transistors no longer exist. Early point contact and junction devices were, for example, often ruined if dropped. Today, both germanium and silicon transistors are quite rugged and this kind of damage is practically a thing of the past.

10.1 Cutters

Transistors can be ruined by cutting the leads with diagonal cutters that use a crimping action instead of a shearing motion similar to scissors. The force with which the cut lead flies through the air gives an indication of the shock wave that travels up the lead into the semiconductor material. Modern transistors can withstand considerable punishment of this type, but it is wise to use a scissor-like cutting tool or else hold the transistor lead near the case with a pair of pliers. If the transistor is already soldered to a printed-circuit board, the soldered connections will, of course, dampen any shock waves traveling toward the device.

10.2 Epoxy Cases

A relatively new type of package that deserves special mention is the epoxy case transistor. These cases are usually used with the less expensive line of silicon devices;

however, many of these transistors have surprisingly good specifications. Cost is quite reasonable. When these units were first introduced, there was some apprehension due to the moisture absorbing tendencies of epoxy. Time and experience, however, have shown these fears to be unjustified. The only place where leakage from moisture has been a problem is on the surface of the case. The leads are usually very close together, and on some units they extend out of both the top and bottom of the package. Heat dissipation is not as good on epoxy units as on conventional metal-case transistors and clip-on heat sinks may, in some cases, prove worthwhile.

10.3 Silicon Grease

When replacing power transistors, it is extremely important to use silicon grease between the transistor and heat sink. Heat is the number one destroyer of transistors, with germanium being more sensitive than silicon. Between the transistor and the surrounding ambient area there are three areas of thermal resistance; namely, between the junction and case, between the case and heat sink, and between the heat sink and ambient area. Silicon grease greatly reduces the thermal resistance between the case and heat sink and this makes it easier for the heat to move from the semiconductor chip to the outside world.

10.4 Dip Soldering

Special precautions must be taken when dip-soldering transistors that are mounted on printed-circuit boards. When the solder is in the molten state, the transistor leads may slide down through the holes in the printed circuit boards. With conventional transistor packages, the metal case may, as a result of the slippage, short out two or more of the printed-circuit conductors. This problem is solved by inserting a small insulating spacer between the transistor and the printed-circuit board. Another solution is to put an S-shaped crimp in the transistor lead. When the transistor is mounted, part of the S is below the board and part is above and this locks the transistor in position.

CHAPTER 11

Troubleshooting Techniques

11.1 General Procedures

To function as an amplifier, a transistor must have its emitter-base junction forward biased. Figures 11-1A and B show typical circuit arrangements for normal forward bias conditions. Thus, the first thing to do is to see if the transistor base-emitter junction is forward biased. Generally, if a VTVM is used, it is best to measure the emitter and base voltages with respect to ground, since ground

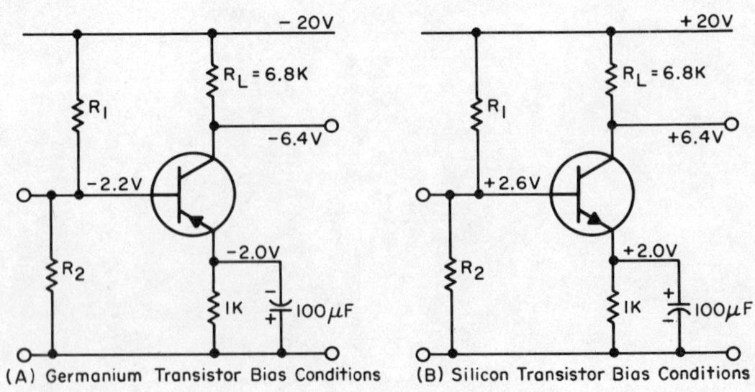

Fig. 11-1. Transistor test circuits.

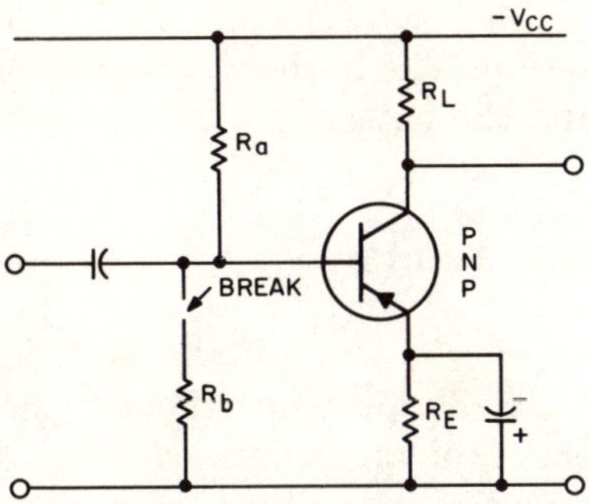

Fig. 11-2. Modifying a circuit for troubleshooting purposes.

loop currents of sufficient amplitude to damage the transistor may be flowing through the VTVM. Many new VTVM's have exceptional high isolation and can be placed directly across the base-emitter junction. Older VTVM's, however, may damage a transistor if placed directly between the emitter and base. It is best to check the characteristics of a VTVM before it is used.

If the transistor is forward biased, the next thing to do is to see if it is *behaving as an amplifier*. Short the emitter to the base to remove the forward bias. The collector voltage should then rise to the approximate level of the supply

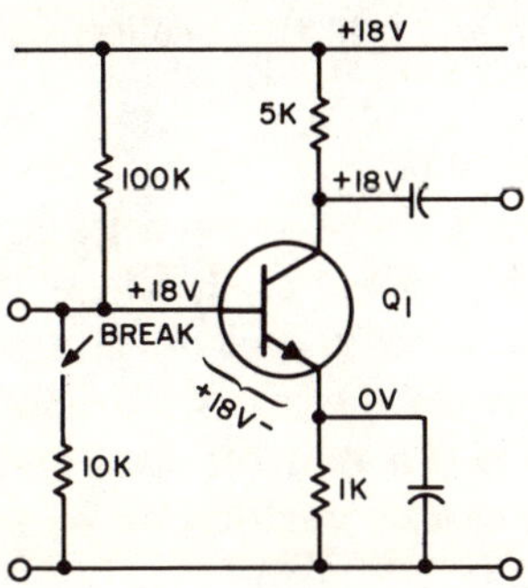

Fig. 11-3. Open emitter.

voltage. (Any difference is caused by I_{CO}, the collector-to-base leakage current. The higher the collector voltage rises, the lower I_{CO}, and the better the transistor.)

11.2 Transistor Faults

A few voltage checks are usually sufficient to reveal a defective transistor. To pinpoint the specific cause of failure, refer to the typical circuit of Fig. 11-2 and open resistor R_b. If there is no voltage drop across resistor R_a, Fig. 11-3, the emitter is open. The absence of a voltage drop across resistor R_L and the presence of a voltage drop across R_E indicates an open collector. See Fig. 11-4. The

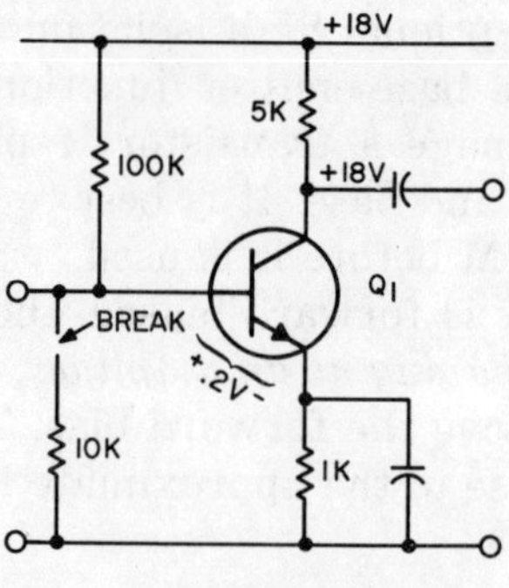

Fig. 11-4. Open collector.

absence of forward bias, Fig. 11-5, indicates an emitter-to-base short. Finally, if there is no voltage drop between the emitter and collector, Fig. 11-6, these two electrodes are short circuited. Be careful not to confuse the last noted symptom with a saturated transistor having, perhaps, 0.1 volt or less between the emitter and collector. The conditions of saturation can be verified by momentarily shorting the emitter and base while monitoring V_{CC}, as previously described.

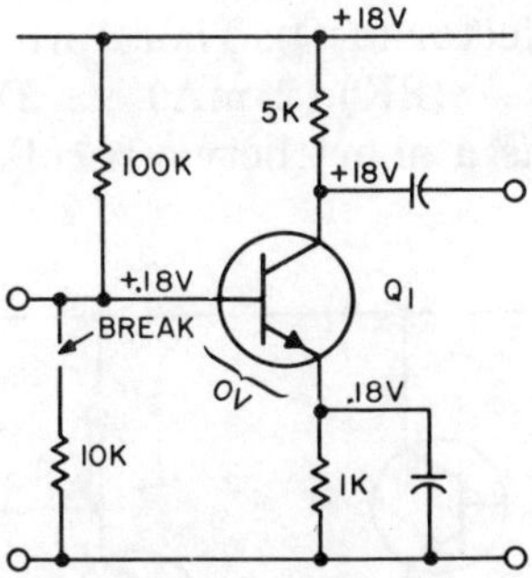

Fig. 11-5. Emitter-to-base short.

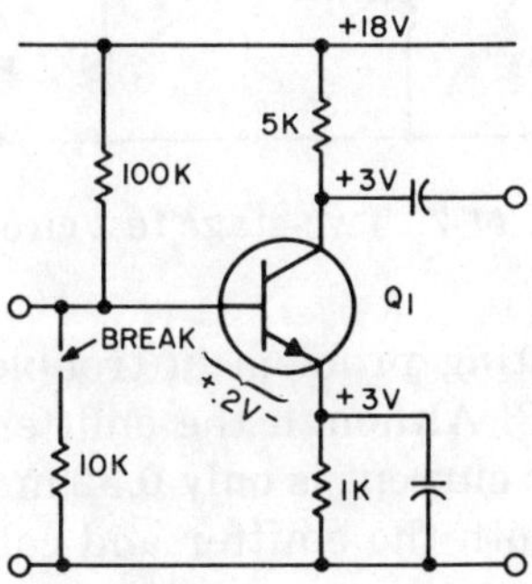

Fig. 11-6. Emitter-to-collector short.

11.3 Other Faults

If the emitter-base junction seems to be reverse biased, look for a shorted coupling capacitor between the emitter of the stage under test and the collector of the previous stage.

An interesting problem is illustrated in Fig. 11-7. In this circuit, both transistors are of the NPN type. Note that Q_2 has 0.8 V *reverse* bias on its emitter-base junction, but the 2.0 volts on the emitter means that there is 2 mA of emitter current. Now, since the emitter-base junction is not shorted, this 2 mA of current also flows through the 8K resistor in the collector of Q_2. Therefore, the collector voltage, V_{CC}, is 18V − (8K) (2 mA) = 2V. Thus, it would appear that Q_2 has a short between collector and emitter.

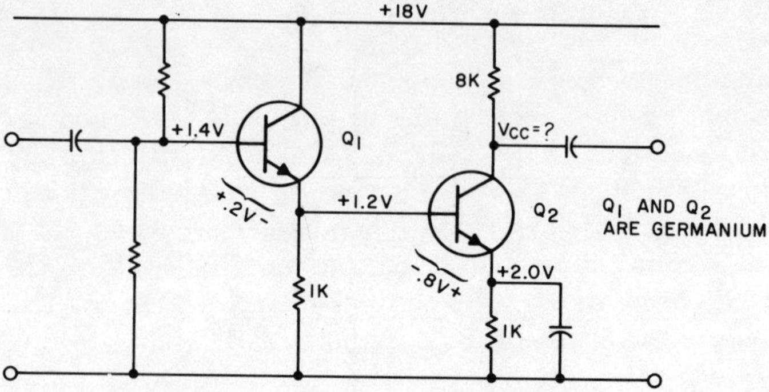

Fig. 11-7. Two-stage test circuit.

Another interesting problem in troubleshooting is illustrated in Fig. 11-8. Although the emitter current of Q_1 is 1 mA, the collector current is only 0.52 mA. Stage Q_2 shows 5 mA flowing in both the emitter and collector circuits, so

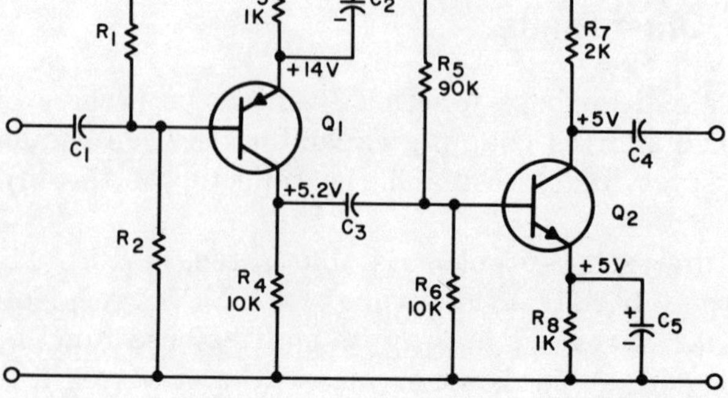

Fig. 11-8. Circuit for practice troubleshooting.

Q_2 is either shorted or saturated. The one voltage that would answer this question is not given; i.e., the voltage on the base of Q_2. If everything were working correctly, this voltage would be approximately

$$V_B = \frac{(15)\,(10K)}{100K} = 1.5 \text{ d-c volts}$$

What appears to have happened is that C_3 is shorted. This would explain why there is only 0.52 mA flowing through resistor R_4. The other 0.48 mA is flowing through C_3 and resistor R_6. If C_3 were shorted, it would also explain the voltages on Q_2, which would then have the correct 0.2 V forward bias on the base-emitter junction. The 5.2 V on the base would produce 5.0 volts on the emitter, which, in turn, would cause the 5 mA of d-c current to flow and Q_2 would saturate.

If capacitor C_3 were replaced, the base voltage of Q_2 would be 1.5 V dc, and the voltage on the emitter would be about 1.3 V dc. This, in turn, would cause about 1.3 mA of dc to flow. The resultant collector voltage would be 12.4 V dc.

This circuit has a very interesting characteristic; i.e., even with C_3 shorted it produces a voltage gain of about 200, and it would be rather easy to assume the circuit is working properly. However, using the fast techniques for voltage gain calculation, this amplifier should produce a gain of:

$$\begin{aligned} A_e &\approx \left(- \frac{0.9R_L}{h_{ib}} \right) \left(- \frac{0.9R_L}{h_{ib}} \right) \\ &\approx \left(\frac{0.9 \times 10K \parallel 10K \parallel 2K}{30} \right) \left(\frac{-0.9 \times 2K}{20} \right) \\ &\approx 4000 \end{aligned}$$

Thus, with a 1 m-V input, the output would be about 4 volts. In other words, most amplification stages without a feedback resistor in the emitter have a gain of around 100 and two stages have a gain of nearly 10,000. This is much greater than a gain of 200.

11.4 Out-of-Circuit Ohmmeter Tests

In about 90 percent of cases, an ohmmeter can be used to test a transistor. However, a certain amount of care must be used as every ohmmeter has a few ranges that put out enough current or voltage to damage the transistor. Table 11-1 describes several of the more popular ohmmeters. The bold face numbers show the completely safe operating ranges. Although the R × 10 and R × 100 ranges are usually not in bold face, they would be safe on all but VHF and fast switching transistors. Since the transistor will have an effective resistance of its own, the actual current flow will be considerably below the values of current given in Table 11-1.

1. *Small signal (PNP Germanium) (see Fig. 11-9A)*
   Positive lead to emitter—common
   Negative lead to base—approximately 200 to 500 ohms
   Negative lead to collector—approximately 10K to 100K
   (Short collector to base and R should be lower than base alone, i.e., shows transistor action)

Fig. 11-9. (A) Ohmmeter readings for a PNP small signal, Ge transistor; (A) ohmmeter readings for a PNP power, Ge transistor; (B) ohmmeter readings for an NPN, small signal, Si transistor; and (B) ohmmeter readings for an NPN, power, Si transistor.

2. *Power (PNP Germanium) (see Fig. 11-9A)*
   Positive lead to emitter—common
   Negative lead to base—approximately 30 to 50 ohms

Negative lead to collector—approximately several hundred ohms

(Short collector to base and R should be lower than base alone, i.e., shows transistor action)

3. With NPN, reverse meter polarity.
4. *Small signal (NPN Silicon) (see Fig. 11-9B)*
   Negative lead to emitter—common
   Positive lead to base—1K to 3K
   Positive lead to collector—very high, may read open.
5. *Power (NPN Silicon) (see Fig. 11-9B)*
   Negative lead to emitter—common
   Positive lead to base—200 ohms to 1K
   Positive lead to collector—high, often greater than 1M

TABLE 11-1. CHARACTERISTICS OF COMMON OHMMETERS

Make, Model, and Range	Open Circuit Voltage	Short Circuit Current	Polarity
HP 412A (VTVM)			
R × 1	0.01 V	8.0 mA	
R × 10	0.1 V	10.0 mA	RED +
R × 100	1.0 V	10.0 mA	
***R × 1K**	**1.0 V**	**1.0 mA**	BLACK −
R × 10K	**1.0 V**	**100.0 μA**	
R × 100K	**1.0 V**	**10.0 μA**	
R × 1M	**1.0 V**	**1.0 μA**	
R × 10M	**1.0 V**	**0.1 μA**	
HP 410B (VTVM)			
R × 1	1.1 V	120 mA	
R × 10	1.1 V	11 mA	RED −
R × 100	**1.1 V**	**1.1 mA**	
R × 1K	**1.1 V**	**110.0 μA**	BLACK +
R × 10K	**1.1 V**	**11.0 μA**	
R × 100K	**1.1 V**	**1.1 μA**	
R × 1M	**1.1 V**	**0.11 μA**	

* Numbers in bold type indicate safe range.

TABLE 11-1. CHARACTERISTICS OF COMMON OHMMETERS (Cont.)

Make, Model, and Range	Open Circuit Voltage	Short Circuit Current	Polarity
HP 410C (VTVM)			
R × 10	1.3 V	55 mA	
R × 100	1.3 V	5.7 mA	RED +
R × 1K	**1.3 V**	**0.57 mA**	BLACK −
R × 10K	**1.3 V**	**57 μA**	
R × 100K	**1.3 V**	**5.7 μA**	
R × 1M	**1.3 V**	**0.5 μA**	
R × 10M	**1.3 V**	**0.05 μA**	
SIMPSON 260 (VOM)			
R × 1	1.5 V	125 mA	RED +
R × 100	**1.5 V**	**1 mA**	
R × 10K	7.5 V	60 μA	BLACK −
SIMPSON 269 (VOM)			
R × 1	1.5 V	74 mA	
R × 10	1.5 V	8 mA	RED −
R × 100	1.5 V	8 mA	BLACK +
R × 1K	**1.5 V**	**0.82 mA**	
R × 10K	24 V	1.3 mA	
R × 100K	30 V	13 μA	
TRIPLETT 630 (VOM)			
R × 1	1.5 V	320 mA	RED −
R × 10	1.5 V	32 mA	BLACK +
R × 100	**1.5 V**	**3.25 mA**	(Varies with
R × 1K	**1.5 V**	**325 μA**	serial
R × 100K	22.5 V	**70 μA**	number)
TRIPLETT 310 (VOM)			
R × 1	1.5 V	7.5 mA	RED −
R × 10	**1.5 V**	**750 μA**	BLACK +
R × 100	**1.5 V**	**75 μA**	(Varies with
R × 10K	1.5 V	75 μA	serial
			number)

* Numbers in bold type indicate safe range.

Ohmmeters have limitations in transistor testing. It is difficult, for example, to identify a power transistor with high leakage current.

If you are working with transistors a great deal, it is probably advisable to buy one of the commercial transistor testers. A transistor checker should at least test for leakage and give h_{FE} or h_{fe} at some current level such as 1 mA.

11.5 Rapid Troubleshooting Procedures

Taking all the information we have covered and orienting it specifically toward rapid troubleshooting techniques lead to a logical summary of practical tips. Before decribing specific tips let's take a moment and review the key transistor characteristics that are important from a troubleshooting point of view.

A conventional PNP or NPN transistor has three operating states:

A. *Off*, that is an open switch.

B. *Part way on*, bias voltages are set so the transistor can amplify, i.e., it can be turned further on or further off. This is the normal bias condition for amplifiers.

C. *Saturated*, behaves like a closed switch. Saturation is defined as where the IR drop across the emitter and collector resistors equals the supply voltage. The interesting thing about saturation is that both the base-emitter and base-collector diodes are forward biased. A saturated germanium transistor may have as low as 0.05 volts between its emitter and collector while a saturated silicon transistor might have about 0.5 volts between these leads. *Saturated* or *off* are the usual conditions found in digital circuits.

In troubleshooting transistor circuits, the most important area to examine is the base-emitter junction as this is the control point of the transistor. Remember that conventional PNP and NPN transistors are basically "off" devices and must be biased "on" to their operating point. This is done by forward biasing the base-emitter junction. Therefore, the status of the base-emitter diode tells exactly what the transistor should be doing. This diode is made out of either silicon or germanium. If the transistor is silicon and has approximately 0.6V forward bias between base and emitter, the transistor should be "on." The amount it should be on depends upon the current gain (β) of the transistor, the resistors in series with the collector and emitter, and the supply voltage. If the transistor were germanium and had approximately 0.2 volts forward bias between base and emitter, it would behave in the same general fashion.

If the transistor has zero bias or reverse bias on its base-emitter junction, it should be turned off. If it is not off under these conditions, it is either shorted or leaky.

This review leads us to our first troubleshooting tip.

TIP #1: Measure the base-emitter voltage. From this decide how the transistor should be behaving. Then look at the collector voltage and see if the transistor is behaving as it should be.

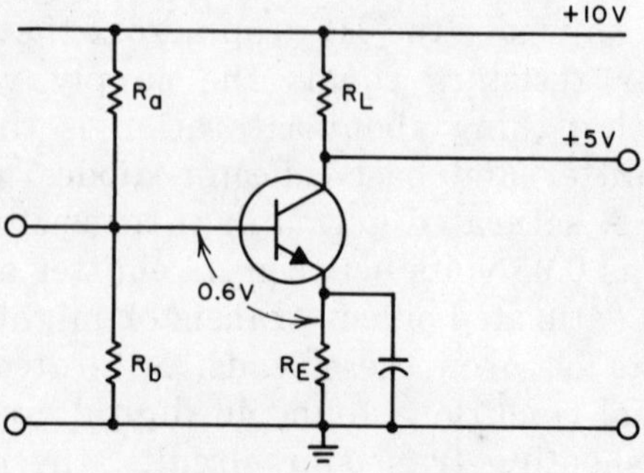

Fig. 11-10. Properly biased transistor.

For example, if the base-emitter voltage is 0.6V forward biased and the collector voltage is the same as the supply voltage, something is wrong. Probably the collector-base junction is open.

Expanding on this idea leads to our second troubleshooting tip.

TIP #2: Modify the control signals present and see if the circuit responds accordingly.

For example, Fig. 11-10 shows a normally biased NPN silicon transistor with the bias resistors adjusted to have the transistor turned on half way. Now remove the forward bias on the base-emitter diode junction by adding

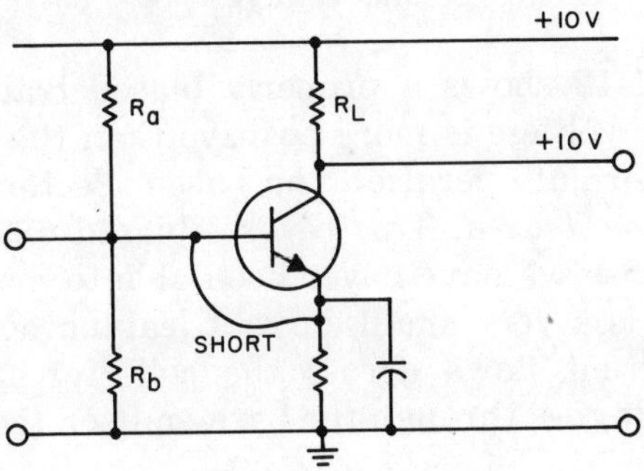

Fig. 11-11. Amplifier with forward bias removed.

the short as is shown in Fig. 11-11. When the short is added, the collector voltage should rise to within a few tenths of a volt of the supply voltage. If it doesn't, we've identified a bad transistor. This technique is perfectly safe in a-c coupled circuits. In some d-c coupled circuits we could cause damage if base-emitter shorts are applied around high power levels such as the output stage of a power amplifier.

When we use this technique, the collector voltage would rise to exactly the supply voltage if there was collector-base leakage current. Since all PNP and NPN transistors have some leakage let's review this area.

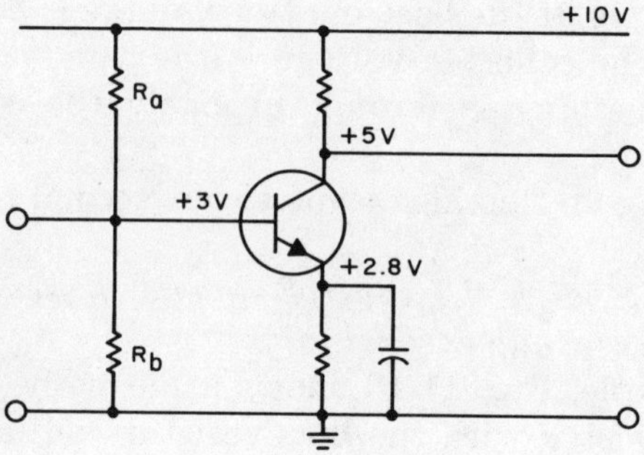

Fig. 11-12. NPN germanium transistor with bias voltages.

Figure 11-12 shows a properly biased transistor. Note the collector voltage is more positive than the base voltage —thus in normal operation the base-collector diode junction is *reverse biased.* This reverse-biased diode should be off but because we have never been able to make a perfect diode there is a very small current leaking across it. This leakage current flows across the collector-base junction and part of it goes through the base-emitter (control point) junction.

Since leakage current is extremely temperature sensitive we can use this to our advantage in troubleshooting:

TIP #3: When a transistor with excessive leakage is sprayed with coolant, it often starts behaving properly. Conversely, heating a leaky transistor will make the problem much worse.

In an amplifier stage excessive leakage current can cause clipping distortion because of the shift in the quiescent operating point.

TIP #4: In an amplifier with clipping distortion try cooling each transistor. Quite likely you will find that when one transistor is cooled the clipping distortion disappears. That transistor probably has excessive leakage.

Even though all these tips are good ones there is a transistor tester that will speed up troubleshooting even more.

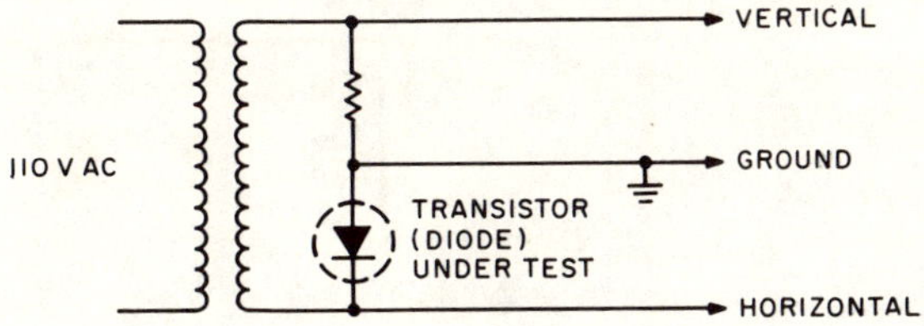

Fig. 11-13. Transistor checker (simplified schematic).

This tester works on the known fact that PNP and NPN transistors are made up of two diodes and examines each diode independently. The display is shown on an oscilloscope.

Figure 11-13 shows a simplified schematic of the transistor checker. This schematic shows that the scope vertical signal is proportional to the current *through* the transistor diode while the horizontal signal is proportional to the voltage *across* the transistor diode.

With the tester connected as shown we would expect the following waveforms:

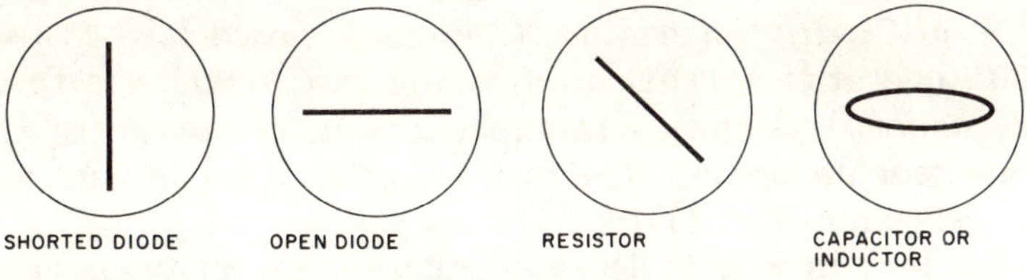

Fig. 11-14. Tester waveforms.

Since our transistor checker puts out a sine wave that has alternatively positive and negative half cycles we would expect a perfect diode to behave as shown in Fig. 11-15.

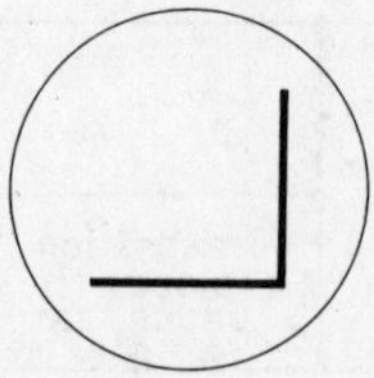

Fig. 11-15. Waveform of good diode.

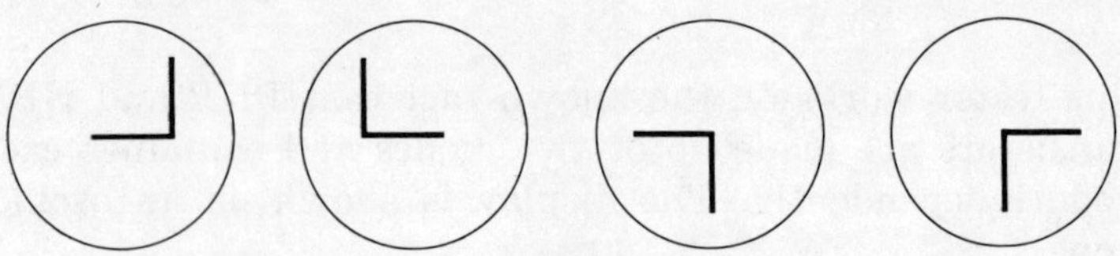

Fig. 11-16. Possible waveforms for a good transistor diode junction.

In actual practice the waveforms shown in Fig. 11-16 are obtained because we do not care which lead is on the base and which lead is on the collector (or emitter).

The waveforms in Fig. 11-16 are typical of *out-of-circuit* transistor checks. Note in Fig. 11-17, which shows a complete schematic, there is a switch for "*In-Circuit*" and "*Out-of-Circuit*" operation. (Switch is closed for "*In-Circuit*" operation.) When performing *In-Circuit* tests there are usually resistors and capacitors associated with the transistor under test. The result is often a waveform such as is shown in Fig. 11-18.

The loop in Fig. 11-18 shows there is associated capacitance (probably a coupling capacitor) and the fact that the waveform is not a perfect "right" angle is because of the associated resistance (probably bias or load resistors).

This transistor tester leads to our next troubleshooting tip.

TIP #5: Use the transistor checker for rapid testing. Make sure to test both the base-emitter and base-collector diodes.

A little experimenting with a printed circuit board containing many transistors will rapidly show you the various waveforms you will encounter for good transistors. The important thing to look for is whether or not the waveform has a "break" in it (pt A in Fig. 11-18). If it does, the transistor diode is good. Remember, the lower the bias resistors, the less defined the "break" (pt A Fig. 11-18) and the more the waveform appears like a "short." Of course, when testing out-of-circuit the "break" will be very sharp—just like a true diode.

This testor can also be used for testing tunnel diodes. The waveform is shown in Fig. 11-19.

When testing tunnel diodes, make sure the switch is in the *In-Circuit* position as you need the extra current.

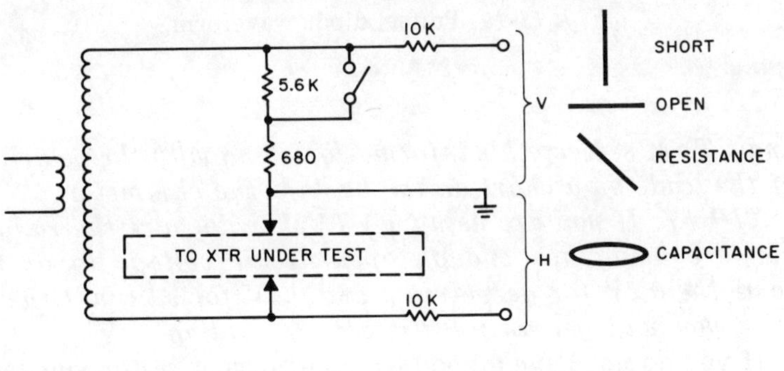

Fig. 11-17. Transistor checker (complete schematic).

Section 11.4 described using an ohmmeter to test a transistor. From this information we obtain tips 6 and 7.

TIP #6: Measure the short-circuit current and open-circuit voltage for each resistance scale on your VOM's

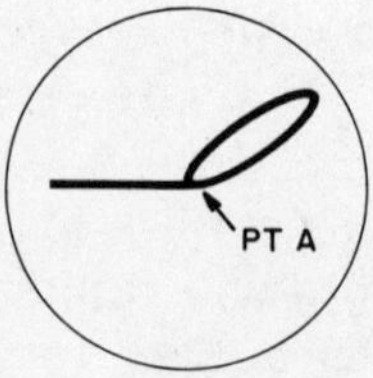

Fig. 11-18. Typical in-circuit waveform for a good transistor.

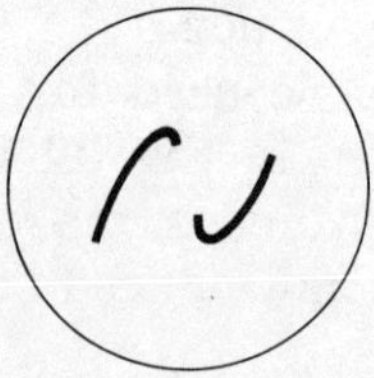

Fig. 11-19. Tunnel diode waveform.

and VTVM's. Keep this information along with the polarity of the leads on a chart on the back of the ohmmeter.

TIP #7: If you are using a VTVM make sure the range you are using has enough open-circuit voltage to overcome the 0.2V for germanium and 0.6V for silicon. Otherwise you will get an unsatisfactory reading.

If you do not have a good leakage current tester you can still measure I_{CBO} with the arrangement shown in Fig. 11-20. Short the emitter-base junction simultaneously measuring the drop across the collector load resistor. For example, if you did this and measured 30 mv across a 10K load resistor (with the emitter shorted) your leakage current would be $I = E/R = 30mv/10K$ or $3\mu a$, which would be about right for a germanium transistor at room tem-

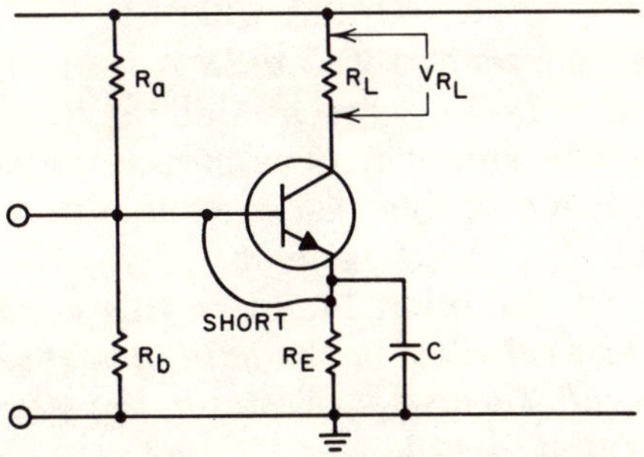

Fig. 11-20. Technique for measuring I_{CBO}.

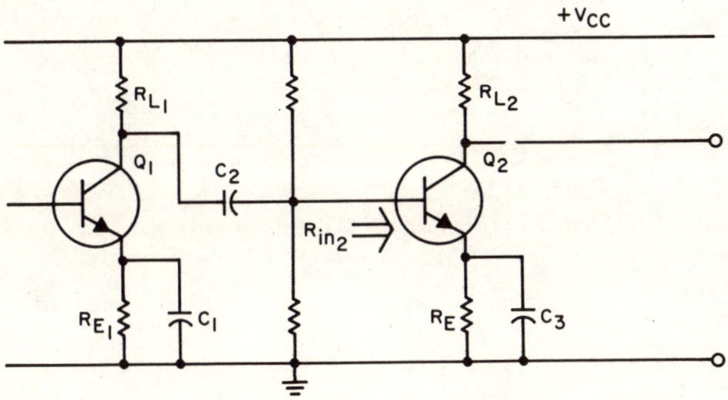

Fig. 11-21. Two-stage amplifier.

perature but is a little high for a silicon surface-passivated transistor.

TIP #8: Measure I_{CBO} by shorting the emitter-base junction and monitoring the voltage across the collector load resistor. $I_{CBO} = V_{RL}/R_L$ (see text).

One of the most common mistakes in analyzing transistor circuits is to miscalculate a stage gain in a multistage amplifier. For example, an excellent approximation of stage gain is $A_e \approx -R_L/h_{ib}$ where h_{ib} is 30Ω at 1MA of d-c emitter current, 15Ω at 2MA, etc. The problem comes in plugging in the correct value for R_L. Figure 11-21 shows a two-stage amplifier. The correct value for R_{L_1} is not the actual value of this resistor but rather the parallel combination of R_{L_1}, R_a, R_b, and R_{in} of Q_2. Usually the R_{in} of Q_2 is the most dominant factor in this combination.

TIP #9: When calculating the gain of a stage be sure to include the parallel loading effects of the next stage bias resistors and input impedance.

All of these tips relate back to important characteristics of transistors. Of course, there are many other tips that are common to NPN and PNP transistors as well as to FET's and vacuum tubes but that is beyond the scope of this book.

APPENDIX 1

Rule-of-Thumb Formulas

Grounded Emitter ($R_L < 20K$ and neglects biasing resistors. R_o is the R_o of the circuit):

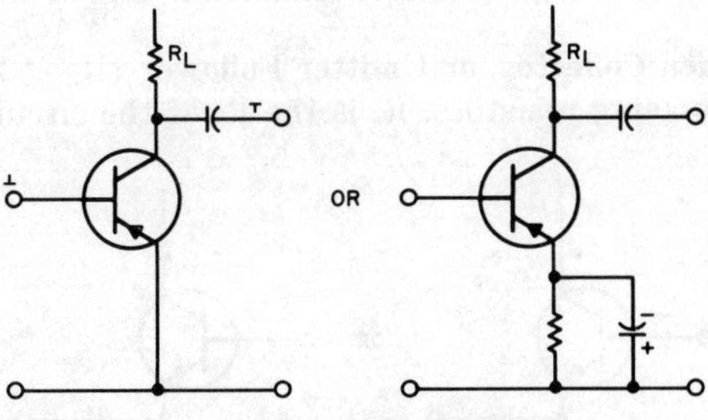

⊥ = POSITIVE GOING SIGNAL
T = NEGATIVE GOING SIGNAL

$$A_e \approx \frac{-0.9R_L}{h_{ib}} \approx -0.03R_L \mid I_E \text{ (mA)} \mid$$

$$A_i \approx 0.8h_{fe} \text{ or } 0.8\beta$$

$$R_i \approx 0.8h_{ie} \approx 0.8h_{fe}h_{ib} \approx \frac{24h_{fe}}{\mid I_E \text{ (mA)} \mid}$$

$$R_o \approx R_L \text{ since } R_o \text{ of transistor} \approx \frac{1}{h_{oe}}$$

Emitter Feedback ($R_L < 20K$ and neglects biasing resistors. R_o is the R_o of the circuit):

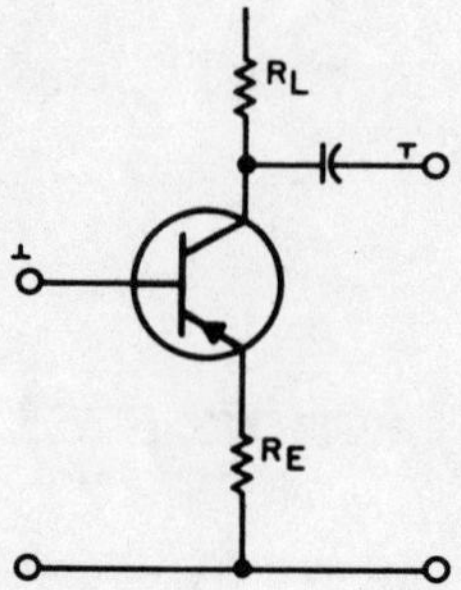

$$A_e \approx \frac{-R_L}{R_E + h_{ib}} \approx \frac{-R_L}{R_E} \text{ if } R_E >> h_{ib}$$

$$A_i \approx 0.8h_{fe}$$

$$R_i \approx 0.8h_{fe}(R_E + h_{ib})$$

$$0.8h_{fe}R_E \qquad \text{if } R_E \geqq 500\Omega$$

$$R_o \approx R_L \text{ since } R_o \text{ of transistor} \approx \frac{h_{fe}}{3h_{oe}}$$

Grounded Collector, or Emitter Follower ($R_L < 20K$ and neglects biasing resistors. R_o is the R_o of the circuit):

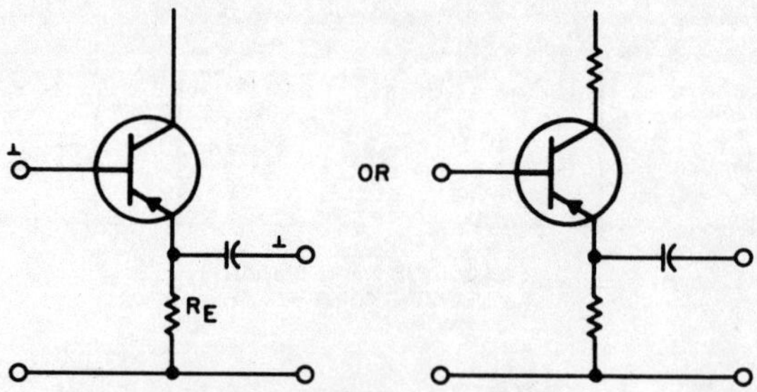

$$A_e \approx \frac{+R_E}{R_E + h_{ib}} \approx +1 \qquad \text{if } R_E >> h_{ib}$$

$$A_i \approx -0.8h_{fe} \text{ or } -0.8\beta$$

$$R_i \approx 0.8h_{fe}\,(R_E + h_{ib})$$

$$0.8h_{fe}\,R_E \qquad \text{if } R_E \geqq 500\Omega$$

$$R_o \approx \frac{h_{ie} + R_g}{h_{fe}} \approx 10 \text{ to } 500\Omega$$

Collector Feedback ($R_L < 20K$ and neglects biasing resistors. R_o is the R_o of the circuit):

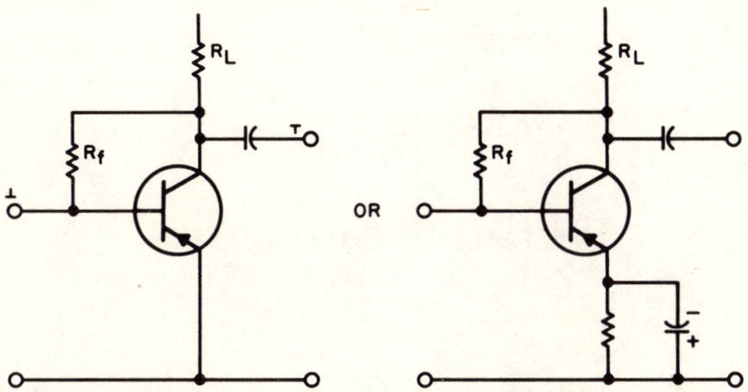

$$A_e \approx \frac{-R_L \parallel R_f}{h_{ib}} \approx -0.03R_L \mid I_E\ (mA) \mid$$

$$A_i \approx \frac{R_f}{R_L}$$

$$R_i \approx \frac{R_f\, h_{ib}}{R_L \parallel R_f} \qquad \text{if } R_f >> R_L$$

$$R_i \approx h_{ib} \qquad \text{if } R_f << R_L$$

$$R_i \approx 2h_{ib} \qquad \text{if } R_f \approx R_L$$

$$R_o \approx R_f \left(\frac{h_{ib}}{R_g} + \frac{1}{h_{fe}}\right) \parallel R_L \qquad \text{if } R_f >> R_g$$

$$R_o \approx \left(\frac{R_f}{h_{fe}} + h_{ib}\right) \parallel R_L \qquad \text{if } R_g >> R_f$$

APPENDIX 2

Exact Formulas

Grounded Emitter (R_o is the R_o of the device and not the circuit. $\triangle = h_{ie}h_{oe} - h_{re}h_{fe}$):

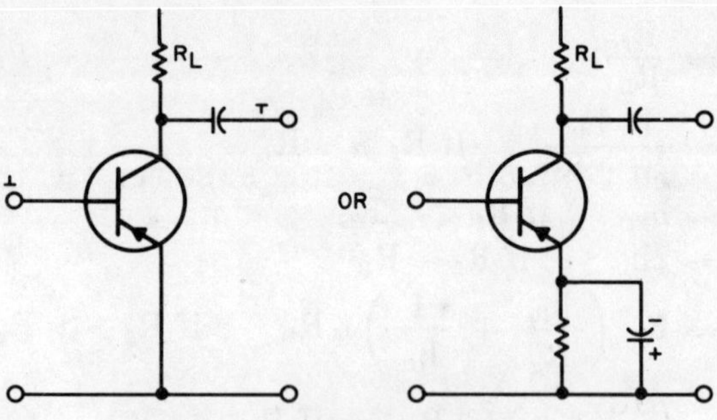

$$A_i = \frac{h_{fe}}{1 + h_{oe}R_L}$$

$$R_i = h_{ie} - A_i\, R_L h_{re}$$

$$A_e = -\, A_i\, \frac{R_L}{R_i}$$

$$R_o = \frac{h_{ie} + R_g}{\triangle + h_{oe}R_g}$$

Emitter Feedback (R_o is the R_o of the device and not the circuit):

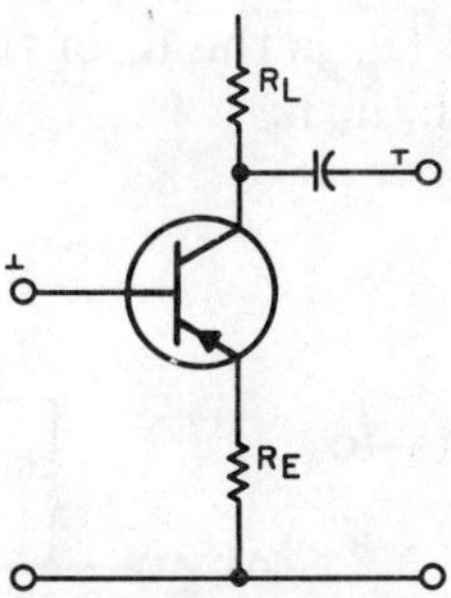

$$A_i = \frac{h_{fe} - h_{oe}R_E}{1 + h_{oe}(R_L + R_E)}$$

$$R_i = h_{ie} + (1 + A_i)R_E - \frac{h_{re}}{h_{oe}}(h_{fe} - A_i)$$

$$A_e = -A_i \frac{R_L}{R_i}$$

$$R_o = R_E + \frac{1}{h_{oe}} + \frac{\left(\frac{h_{fe}}{h_{oe}} - R_E\right)\left(R_E + \frac{h_{re}}{h_{oe}}\right)}{R_g + h_{ie} + R_E - \frac{h_{re}h_{fe}}{h_{oe}}}$$

Grounded Collector or Emitter Follower (R_o is the R_o of the device and not the circuit):

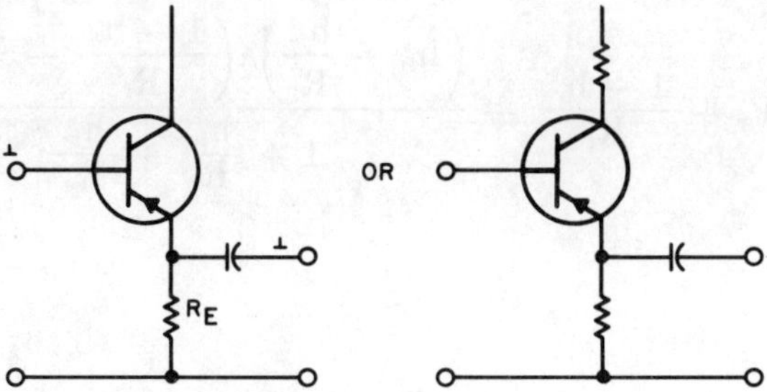

$$A_i = -\frac{1 + h_{fe}}{1 + h_{oe}R_E}$$

$$R_i = h_{ie} + (h_{re} - 1)\ A_i\ R_E$$

$$A_e = -A_i \frac{R_E}{R_i}$$

$$R_o = \frac{R_g + h_{ie}}{(R_g + h_{ie})h_{oe} + (h_{fe} + 1)(1 - h_{re})}$$

Collector Feedback (R_o is the R_o of the device and not the circuit. $\triangle = h_{ie}h_{oe} - h_{re}h_{fe}$) :

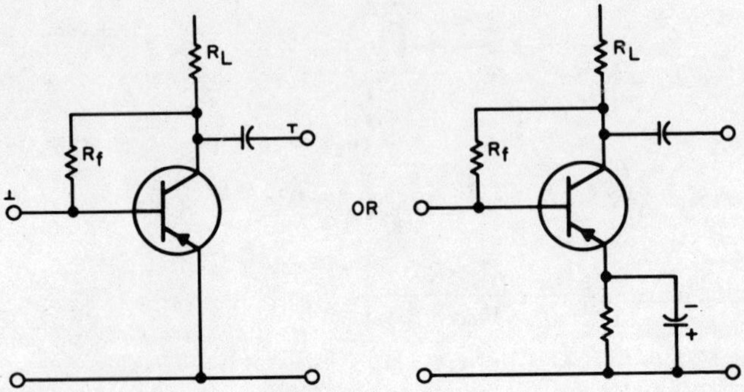

$$A_i = \frac{R_f h_{fe} - h_{ie}}{h_{ie} + R_f + h_{oe}R_L R_f + R_L\,(1 + h_{fe} + \triangle - h_{re})}$$

$$R_i = \frac{\triangle + \dfrac{h_{ie}}{R_L \parallel R_f}}{h_{oe} + \dfrac{h_{ie}}{R_L R_f} + \dfrac{1}{R_L} + \dfrac{1 + h_{fe} - h_{re} + \triangle}{R_f}}$$

$$A_e = -\,A_i\,\frac{R_L}{R_i}$$

$$R_o = \left[\, h_{oe} + \frac{1 - h_{re}}{R_f} + \frac{\left(h_{fe} - \dfrac{h_{ie}}{R_f}\right)\left(\dfrac{1 - h_{re}}{R_f} - \dfrac{h_{re}}{R_g}\right)}{1 + \dfrac{h_{ie}}{R_f} + \dfrac{h_{ie}}{R_g}} \,\right]^{-1}$$

Index